Life Science

____ Grade 1 ____

Written by Tracy Bellaire

The experiments in this book fall under ten topics that relate to two aspects of life science: **Needs and Characteristics of Living Things; and Exploring the Senses.** In each section you will find teacher notes designed to provide you guidance with the learning intention, the success criteria, materials needed, a lesson outline, as well as provide some insight on what results to expect when the experiments are conducted. Suggestions for differentiation are also included so that all students can be successful in the learning environment.

Tracy Bellaire is an experienced teacher who continues to be involved in various levels of education in her role as Differentiated Learning Resource Teacher in an elementary school in Ontario. She enjoys creating educational materials for all types of learners, and providing tools for teachers to further develop their skill set in the classroom. She hopes that these lessons help all to discover their love of science!

Published in Canada by:
On The Mark Press
Belleville, ON
www.onthemarkpress.com

Funded by the
Government
of Canada

Canadä

OTM2160   ISBN: 9781487710224
© On The Mark Press

At A Glance

Learning Intentions

	In Your Environment	In the Animal Kingdom	The Plant World	The Human Factor	Needs Intertwined!	Explore Your Senses	Applying Your Senses	Animal Senses	Protecting the Senses	The Human Body
Knowledge and Understanding Content										
Identify things existing that are in their environment, classify them as living or non-living	•									
Discuss, compare, and sort physical characteristics of a variety of animals, research an animal of choice		•								
Identify the main physical characteristics of plants and explain how plants use them to meet their needs			•							
Recognize physical characteristics of a human; identify and compare the basic needs of a human to other living things				•						
Recognize what living things provide for each other; describe the problems that could result from the loss of some kinds of living things					•					
Identify the five senses and recognize how they are used to identify objects in our world						•				
Demonstrate how the senses are applied to describe and sort things in our environment							•			
Research and describe how animals use their senses to survive in the natural world								•		
Identify the protective parts of the eye and ear; and recognize the importance of protecting our senses that contribute to safety in daily living activities									•	
Determine the location and function of major parts in the human body and explain how they are used to meet our needs and explore our world										•
Thinking Skills and Investigation Process										
Make predictions, formulate questions, and plan an investigation		•	•		•					
Gather and record observations and findings using drawings, tables, written descriptions	•	•	•	•	•	•	•	•	•	•
Recognize and apply safety procedures in the classroom	•	•	•	•	•	•	•	•	•	•
Communication										
Communicate the procedure and conclusions of investigations using demonstrations, drawings, and oral or written descriptions, with use of science and technology vocabulary	•	•	•	•	•	•	•	•	•	•
Application of Knowledge and Skills to Society and the Environment										
Create an action plan to help maintain a healthy environment for living and non-living things	•								•	
Describe problems that could result from the loss of some kinds of living things, and who would be affected in certain ways					•					

OTM2160   ISBN: 9781487710224

TABLE OF CONTENTS

OTM2160   ISBN: 9781487710224
© On The Mark Press

Teacher Assessment Rubric

Student's Name: _____   Date: _____

Success Criteria	Level 1	Level 2	Level 3	Level 4
Knowledge and Understanding Content				
Demonstrate an understanding of the concepts, ideas, terminology definitions, procedures and the safe use of equipment and materials	Demonstrates limited knowledge and understanding of the content	Demonstrates some knowledge and understanding of the content	Demonstrates considerable knowledge and understanding of the content	Demonstrates thorough knowledge and understanding of the content
Thinking Skills and Investigation Process				
Develop hypothesis, formulate questions, select strategies, plan an investigation	Uses planning and critical thinking skills with limited effectiveness	Uses planning and critical thinking skills with some effectiveness	Uses planning and critical thinking skills with considerable effectiveness	Uses planning and critical thinking skills with a high degree of effectiveness
Gather and record data, and make observations, using safety equipment	Uses investigative processing skills with limited effectiveness	Uses investigative processing skills with some effectiveness	Uses investigative processing skills with considerable effectiveness	Uses investigative processing skills with a high degree of effectiveness
Communication				
Organize and communicate ideas and information in oral, visual, and/or written forms	Organizes and communicates ideas and information with limited effectiveness	Organizes and communicates ideas and information with some effectiveness	Organizes and communicates ideas and information with considerable effectiveness	Organizes and communicates ideas and information with a high degree of effectiveness
Use science and technology vocabulary in the communication of ideas and information	Uses vocabulary and terminology with limited effectiveness	Uses vocabulary and terminology with some effectiveness	Uses vocabulary and terminology with considerable effectiveness	Uses vocabulary and terminology with a high degree of effectiveness
Application of Knowledge and Skills to Society and Environment				
Apply knowledge and skills to make connections between science and technology to society and the environment	Makes connections with limited effectiveness	Makes connections with some effectiveness	Makes connections with considerable effectiveness	Makes connections with a high degree of effectiveness
Propose action plans to address problems relating to science and technology, society, and environment	Proposes action plans with limited effectiveness	Proposes action plans with some effectiveness	Proposes action plans with considerable effectiveness	Proposes action plans with a high degree of effectiveness

OTM2160   ISBN: 9781487710224
© On The Mark Press

Student Self Assessment Rubric

Name: _____ Date: _____

Put a check mark ✔ in the box that best describes you:

	Always	Almost Always	Sometimes	Needs Improvement
I am a good listener.				
I followed the directions.				
I stayed on task and finished on time.				
I remembered safety.				
My writing is neat.				
My pictures are neat and colored.				
I reported the results of my experiment.				
I discussed the results of my experiment.				
I know what I am good at.				
I know what I need to work on.				

1. I liked _____

_____

2. I learned _____

_____

3. I want to learn more about _____

_____

INTRODUCTION

The activities in this book have two intentions: to teach concepts related to life science and to provide students the opportunity to apply necessary skills needed for mastery of science and technology curriculum objectives.

Throughout the experiments, the scientific method is used. The scientific method is an investigative process which follows five steps to guide students to discover if evidence supports a hypothesis.

1. **Consider a question to investigate.**
   For each experiment, a question is provided for students to consider. For example, "What makes living things different from non-living things?"

2. **Predict what you think will happen.**
   A hypothesis is an educated guess about the answer to the question being investigated. For example, "I believe that living things need air, water, food, and shelter to live. Non-living things exist without needs". A group discussion is ideal at this point.

3. **Create a plan or procedure to investigate the hypothesis.**
   The plan will include a list of materials and a list of steps to follow. It forms the "experiment".

4. **Record all the observations of the investigation.**
   Results may be recorded in written, table, or picture form.

5. **Draw a conclusion.**
   Do the results support the hypothesis? Encourage students to share their conclusions with their classmates, or in a large group discussion format.

The experiments in this book fall under ten topics that relate to two aspects of life science: **Needs and Characteristics of Living Things and Exploring the Senses.** In each section you will find teacher notes designed to provide you guidance with the learning intention, the success criteria, materials needed, a lesson outline, as well as provide some insight on what results to expect when the experiments are conducted. Suggestions for differentiation are also included so that all students can be successful in the learning environment.

ASSESSMENT AND EVALUATION:

Students can complete the Student Self-Assessment Rubric in order to determine their own strengths and areas for improvement. Assessment can be determined by observation of student participation in the investigation process. The classroom teacher can refer to the Teacher Assessment Rubric and complete it for each student to determine if the success criteria outlined in the lesson plan has been achieved. Determining an overall level of success for evaluation purposes can be done by viewing each student's rubric to see what level of achievement predominantly appears throughout the rubric.

OTM2160   ISBN: 9781487710224
© On The Mark Press

IN YOUR ENVIRONMENT

LEARNING INTENTION:

Students will learn about living and non-living things that are in their environment and the importance of showing care for their environment.

SUCCESS CRITERIA:

- identify things existing in your environment
- classify things as living or non-living, describing the difference
- recognize actions that show care for your school, your home, and your community
- create a personal action plan to keep your environment healthy

MATERIALS NEEDED:

- a copy of "What's In Your Environment?" worksheet 1, 2, 3, and 4 for each student
- a copy of "Helping the Environment" worksheet 5 for each student
- a copy of "How Will You Help Mother Earth?" worksheet 6 for each student
- a copy of "Spread the Word!" worksheet 7 for each student
- chart paper, markers, pencil crayons, clipboards, pencils

PROCEDURE:

*This lesson can be done as one long lesson, or done in two or three shorter lessons.

1. Engage students in a "knee to knee, eye to eye" activity where they turn and talk with a partner to brainstorm a definition of 'environment'. Come back as a large group, record students' ideas on chart paper. Discuss the concept of 'what is environment?' to ensure all students' understanding and to reach a consensus of its meaning.

2. Give students worksheet 1, a clipboard and a pencil. Take them outside to the school yard to look for things that are in their environment.

3. Give students worksheet 2, a clipboard and a pencil. Take them outside to a local park or forested/ wild area to look for things that are in their environment.

4. Upon returning to the classroom, divide students into pairs and give each pair worksheet 3. They will share with their partner three things that they found to be in their environment. They will also engage in a 'Think-Pair-Share' activity to discuss the definition of a living thing, of a non-living thing, and explain the differences. Come back together as a large group to discuss and record their ideas on chart paper. (Living things show characteristics such as growth, movement, respiration, reproduction, environmental adaptation and response.)

5. Give students worksheet 4. Students will sort the things that they drew on worksheets 1 and 2 into the boxes on worksheet 4, categorizing them as either living or non-living.

6. Engage students in a discussion about the importance of a healthy environment.
   - What happens to living things when their environment is not healthy?
   - What are some ways that people harm other living things?
   - What are some ways that people help other living things?

7. Give students worksheets 5, 6, and 7 to complete.

DIFFERENTIATION:

Slower learners may benefit by omitting the top portion of worksheet 3, and completing the 'think-pair-share' activity as a small group with teacher support to ensure the discussion stays on track and that there is a clear understanding of the differences between the two categories.

For enrichment, faster learners could create a poster that promotes the maintenance of a healthy environment (ideas can be taken from their T-shirt design on worksheet 7). These posters could be displayed around the school.

Name:

What's in *Your* Environment?

Have you ever noticed what is in your environment? Take a walk around your school yard. Look for things that are sharing your environment.

In the box below, draw and label what you see or find.

OTM2160   ISBN: 9781487710224
© On The Mark Press

Let's go a little further. Take a walk to your local park or in a forest. Look for things that are sharing your environment.

In the box below, draw and label what you see or find.

Name:

Share and Compare

You and a partner will share 3 things that you each found in the school yard or in your local park.

Draw and label the 3 things that your partner shares in the box below.

Think Pair Share

With your partner, do some thinking and sharing of ideas about the questions below.

"What is a living thing?"

"What is a non-living thing?"

"What makes living things different from non-living things?"

OTM2160   ISBN: 9781487710224
© On The Mark Press

Sort It Out!

Use what you have learned about living and non-living things. Look back at worksheet 1 and 2. Sort the things that you drew into the "living things" box or into the "non-living things" box.

Living things

Non-living things

Helping the Environment

Show how people in **your school** help to keep the environment healthy.

Show how people in **your community** help keep the environment healthy.

Show how people in **your home** help to keep the environment healthy.

OTM2160   ISBN: 9781487710224
© On The Mark Press

How Will You Help Mother Earth?

You have the power to help! Tell what **you** can do to help keep the environment healthy.

To help keep the environment healthy, I can...

_____

_____

_____

_____

To help keep the environment healthy, I can...

_____

_____

_____

_____

OTM2160 ISBN: 9781487710224
© On The Mark Press

Spread the Word!

You have learned a lot about what is in your environment, and why it is important to all things in it, to keep it a healthy place to be.

Design a T-shirt that promotes keeping our environment healthy.

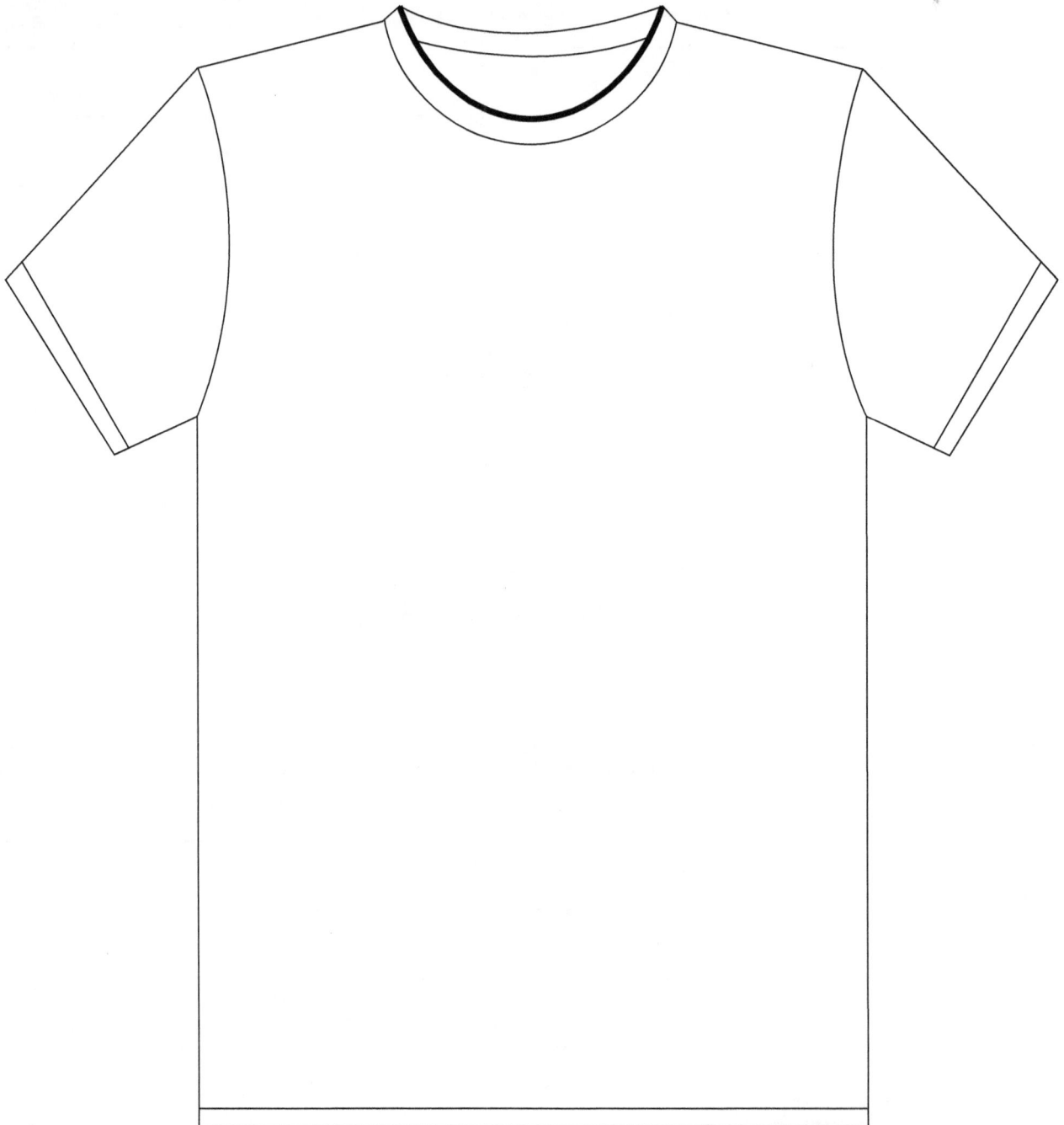

OTM2160   ISBN: 9781487710224

IN THE ANIMAL KINGDOM

LEARNING INTENTION:

Students will learn about the physical characteristics of a variety of animals.

SUCCESS CRITERIA:

- recognize the physical characteristics of a variety of animals
- compare and sort a variety of animals according to their physical characteristics
- research and describe the physical characteristics, habitat, nutrient needs, and special adaptations of an animal
- give reasons for choosing the animal to research

MATERIALS NEEDED:

- a copy of "What Do You Know About Animals?" worksheet 1, 2, 3, and 4 for each group of students
- a copy of "Describing an Animal" worksheet 5 and 6 for each student
- access to the internet, or local library
- Bristol boards (1/2 piece per student)
- chart paper, markers, pencils, pencil crayons, scissors, modeling clay

PROCEDURE:

*This lesson can be done as one long lesson, or done in three or four shorter lessons.

1. Divide students into small groups of 3 or 4. Give each group a copy of worksheets 1, 2, 3, and 4. Instruct them to talk about things that they know about the animals on each of the cards. After students have discussed each animal, ask each student to share one or two things that they know about one or two of the animals on the cards. *An option is to record these student responses on chart paper for reference for future lessons.*

2. Students will continue working within their groups. Give students scissors. They will cut out the animal cards on worksheets 1, 2, 3, and 4. Give students categories for which to sort the animal cards. This will allow them to compare the physical characteristics of the animals, and to discuss why they have them. Some categories are:

   - animals with wings/ no wings—walkers, crawlers, hoppers, flyers, slitherers
   - animals with bills/ no bills—feathers, fur, skin, scales
   - swimmers/ non-swimmers—2 legs, 4 legs, 6 legs, no legs

3. Students will research an animal that they would like to know more about. Give them worksheets 5 and 6. Students can access the internet, or visit the local library to gather information to help them answer the questions on worksheets 5 and 6. Upon completion, students can display their work on Bristol board, which could be posted in the classroom or around the school. *An alternative follow up activity is to have students participate in a carousel activity to read and learn about their classmates' chosen animals.*

DIFFERENTIATION:

Slower learners may benefit by working in a small group with teacher support to complete the research about an animal. A section of the research project could be assigned to each student in the small group, so that all sections are filled in to complete one final project. Each of their sections could be done on large chart paper or on one piece of Bristol board, and then displayed in the classroom.

For enrichment, faster learners could use modeling clay to make a sculpture of their animal.

What Do You Know About Animals?

Group Talk!

With your group members, talk about the things that you know about each animal on the cards below.

	frog		bear
	lizard		dragonfly
	robin		dog
	eagle		elephant
	rattlesnake		bluejay

OTM2160   ISBN: 9781487710224
© On The Mark Press

Name:

With your group members, talk about the things that you know about each animal on the cards below.

	seal		salmon
	butterfly		giraffe
	shark		turtle
	grasshopper		dolphin
	ant		crocodile

OTM2160   ISBN: 9781487710224
© On The Mark Press

Name:

With your group members, talk about the things that you know about each animal on the cards below.

 bee

 whale

 chipmunk

 chicken

 horse

 kangaroo

 cat

 rabbit

 deer

 beaver

OTM2160 ISBN: 9781487710224
© On The Mark Press

Name:

With your group members, talk about the things that you know about each animal on the cards below.

 cow

 pig

 lion

 squirrel

 wolf

 fox

 raccoon

 moose

 duck

 owl

OTM2160   ISBN: 9781487710224
© On The Mark Press

Describing an Animal

Choose an animal that you would like to know more about. Visit a library or use a computer to help you find out the answers to the questions below.

Let's Research!

Animal Name: _____

Where does this animal live? _____

_____

_____

_____

_____

Make an illustration of this animal in its habitat.

OTM2160   ISBN: 9781487710224
© On The Mark Press

Name:

What special adaptations does this animal have?

◆ _____

◆ _____

◆ _____

Why does this animal need these special adaptations?

_____

_____

_____

Make an illustration of this animal in its habitat.

With a partner, talk about why you think your animal is so interesting.

THE PLANT WORLD

LEARNING INTENTION:

Students will learn about the physical characteristics of plants and how they meet their needs.

SUCCESS CRITERIA:

- recognize a variety of plants in your neighborhood
- identify the main physical characteristics of a plant
- compare a variety of plants according to their physical characteristics
- investigate the basic needs of a plant
- explain the function of the main physical characteristics of a plant
- describe how a plant uses its physical characteristics to meet its needs

MATERIALS NEEDED:

- a copy of "Plants in My Neighborhood" worksheet 1 for each student
- a copy of "The Parts of a Plant" worksheet 2 for each student
- a copy of "The Needs of a Plant" worksheet 3, 4, and 5 for each student
- a copy of "The Sum of All Parts" worksheet 6 and 7 for each student
- a copy of "Drinking Up the Nutrients!" worksheet 8, 9, and 10 for each student
- 5 or 6 assorted plants
- 3 small potted plants for each group of students
- medium sized boxes (one for each group of students)
- access to water, a few measuring cups, a couple of watering cans
- a celery stalk with leaves, a tall glass of water, a magnifying glass, a spoon (a set for each pair of students)

- a few small bottles of red food coloring
- chart paper, markers, pencils, pencil crayons, clipboards, labels
- iPods or iPads (optional)

PROCEDURE:

*This lesson can be done as one long lesson, or be done in four or five shorter lessons.

1. Give students worksheet 1, a clipboard and a pencil. Take them out into the neighborhood to look for different kinds of plants. Encourage them to notice the types of trees, shrubs, grasses, flowered plants, etc. that are growing locally. *An option is to give students iPods or iPads to take photos of the different vegetation that they see.*

2. Display 5 or 6 assorted plants (roots exposed). Engage students in a discussion about the physical characteristics of each plant type. How they different? What do they all have in common? Ask students to look back at the plants that they drew on worksheet 1 (or have taken photos of). How are they different? How are they the same? (Some common characteristics that should be noted are that all plants have a root system, stem, leaves. They differ in size, shape, color, and some have flowers). Give students worksheet 2. They will label the basic parts of a plant.

3. Give students worksheets 3, 4, and 5. They will make a prediction about what may happen to a plant that is placed in a dark place, to one that is in sunlight but given no water, and to one that is in sunlight and given water. Over the next two weeks, students will make observations and conclusions about the growth of the plants and record them on worksheet 4 and 5. (Sunlight, water, and air are needed to make plants grow.)

OTM2160   ISBN: 9781487710224

4. Divide students into pairs. Give them worksheet 6. They will engage in a "think-pair-share" activity to discuss and then record answers to the questions on the worksheet. A follow up option is to come back together as a large group to share responses. Student responses could be recorded on chart paper and displayed in the classroom for future reference.

5. Give students worksheet 7. Read through with the students about the main parts of a plant and their purposes. Along with the content, discussion of certain vocabulary words would be of benefit to ensure students' understanding of the concepts.

   Some interesting vocabulary words to focus on are:
   • chlorophyll • attract • nutrients • absorb
   • pollinate • supports • anchor • minerals

6. Divide students into pairs. Give them worksheets 8, 9, 10, and the materials to conduct the experiment. Read through the materials needed and what to do sections to ensure their understanding of the task. Students will make observations before and after the experiment, then make conclusions and a connection to what they have already learned about the needs of plants.

*As an activity to enhance the learning about the physical characteristics and needs of plants, show students The Magic School Bus episode called "Gets Planted". Episodes can be accessed at www.youtube.com

DIFFERENTIATION:

Slower learners may benefit by working with a partner to make and record observations about the plants throughout the two weeks, or simply draw what they saw each week. An additional accommodation would be for these learners to work as one small group with teacher support to discuss the questions on worksheet 6. Answers could be recorded on chart paper and posted in the classroom as reference throughout the unit.

For enrichment, faster learners could plant and care for the 5 or 6 assorted plants (used in item #2). They could chart their growth by measuring the plants' heights, or amount of flower production each has. This could be done by creating a simple bar graph.

Plants in My Neighborhood

Have you ever noticed what kinds of plants are in your neighborhood?

Take a walk around your neighborhood. In the box below, draw and label the plants that you see.

OTM2160 ISBN: 9781487710224
© On The Mark Press

The Parts of a Plant

Label the main parts of the plant below. Use the words in the Word Box to help you.

leaves   roots   stem   flower

Challenge question:

Why are the roots of a plant important?

_____

_____

_____

_____

The Needs of a Plant

Question:

What does a plant need to live and grow?

Materials Needed:

3 small potted plants (one labeled "DARK", one labeled "LIGHT", and one labeled "LIGHT & WATER")

• water	
• watering can	
• measuring cup	
• a medium sized box	

What to do:

1. Make a prediction about the answer to the question. Record it on your worksheet.
2. Place the "**DARK**" plant under the box.
3. Place the "LIGHT" and the "**LIGHT & WATER**" plants in a sunny place.
4. Every 3 days, water the "**DARK**" plant, and water the "**LIGHT & WATER**" plant. Be sure to give them the same amount of water.
5. Record your observations of all plants at the end of each week, for 2 weeks.
6. Make conclusions about what you observed.

OTM2160   ISBN: 9781487710224
© On The Mark Press

Let's Predict

What does a plant need to live and grow?

_____

_____

Let's Investigate

This is what I saw each week.

WEEK 1

	DARK PLANT	LIGHT PLANT	LIGHT & WATER PLANT
What color are the leaves?			
Does the plant look healthy or sick?			

WEEK 2

	DARK PLANT	LIGHT PLANT	LIGHT & WATER PLANT
What color are the leaves?			
Does the plant look healthy or sick?			

Drawing of the plant grown in the dark:

Drawing of the plant grown in sunlight, without water:

Drawing of the watered plant grown in sunlight:

Let's Conclude

What do plants need to live and grow?

_____

_____

How do you know this?

_____

_____

OTM2160   ISBN: 9781487710224
© On The Mark Press

Name:

The Sum of All Parts

Think **Pair** **Share**

With a partner, do some thinking and sharing of ideas about the questions below. Record your ideas.

"What does a plant use its roots for?"

"What does a plant use its stem for?"

"What does a plant use its leaves for?"

"What does a plant use its flowers for?"

OTM2160   ISBN: 9781487710224
© On The Mark Press

Name:

Fast Facts!

Leaves use energy from the sun to mix with their chlorophyll to make a sugar from the air. The sugar, air, and water that the plant takes in, makes food for it.

When flowers are pollinated, they produce fruit. Colorful flower petals attract animals and insects that help to pollinate the flowers. Flowers also make seeds that are needed to start new plants.

A stem of a plant supports it. The stem grows leaves that make food for the plant. The stem stores the food. The stem carries water and nutrients to other parts of the plant.

The roots anchor the plant to the ground. The roots of a plant absorb water and minerals that the plant needs to grow.

OTM2160   ISBN: 9781487710224
© On The Mark Press

Drinking Up the Nutrients!

Do you want to see a plant have a drink of water to get the nutrients it needs to be a healthy plant?

You'll need:

• a celery stalk with its leaves on	
• a tall glass of water	
• a spoon	
• a magnifying glass	
• red food coloring	

What to do:

1. Use the magnifying glass to take a look at the base of the celery stalk. On worksheet 9, describe and draw what you see.

2. Next, put 5 drops of the red food coloring into the glass of water.

3. Stir the water and food coloring with the spoon.

4. Put the celery stalk into the glass of water. Leave it sit over night.

5. The next day, observe the celery stalk. Record your observations on worksheet 9.

6. Make conclusions and a connection about what you have observed. Record them on worksheet 10.

OTM2160   ISBN: 9781487710224
© On The Mark Press

Let's Observe

Before beginning the experiment...

The base of the celery stalk looked

_____

_____

_____

_____

_____

A drawing of the base of the celery stalk:

After doing the experiment...

The base of the celery stalk looked

_____

_____

_____

The stem and leaves looked

_____

_____

_____

A drawing of the celery stalk:

OTM2160   ISBN: 9781487710224
© On The Mark Press

Let's Conclude

Why did the celery stalk and its leaves turn red?

_____

_____

_____

_____

> With a partner, talk about how the structure and design of the celery stalk helps to get its nutrient needs.

Let's Connect It!

If you planted a garden, explain what your plants would need to grow and be healthy.

_____

_____

_____

_____

OTM2160   ISBN: 9781487710224
© On The Mark Press

THE HUMAN FACTOR

LEARNING INTENTION:

Students will learn about the physical characteristics and the basic needs of humans, and compare basic needs to other living things.

SUCCESS CRITERIA:

- identify the physical characteristics of humans and describe each of their uses
- identify common human physical characteristics
- compare differences in human physical characteristics
- identify the four basic needs of a human and describe how these needs are met
- chose an animal, identify its basic needs and describe how they are met
- compare ways in which humans meet their basic needs to that of another living thing
- recognize ways that humans are interfering with the natural world

MATERIALS NEEDED:

- a copy of "Characteristics of the Human Body" worksheet 1 for each group of students
- a copy of "Look at Me, Look at You!" worksheet 2 and 3 for each student
- a copy of "Our Basic Needs" worksheet 4, 5, and 6 for each student
- a copy of "Needing a Comparison" worksheet 7 and 8 for each student
- a copy of "Interfering with Nature" worksheet 9 for each student
- chart paper, markers, pencils, pencil crayons

PROCEDURE:

*This lesson can be done as one long lesson, or be divided into five shorter lessons.

1. Give students worksheet 1. They will identify some human body parts and describe their function. For example, we have 2 arms which function by carrying things, we have a stomach which holds food we eat, we have eyes which we use to see things, etc. A follow up option is to come back together as a large group to hear some students' response, in particular to acknowledge that humans have many parts to their bodies which all have a function, or more than one function.

2. Play a body part matching game. Give students a clue and they will have to guess what part of the body it refers to. Some sample clues are:

   this body part lets you smell
   this body part lets you think
   this body part lets you stand
   this body part lets you bend your legs
   this body part lets you taste
   this body part lets you scratch

3. Divide students into pairs. Give them worksheets 2 and 3. They will illustrate themselves then identify the physical characteristics that their human bodies have in common, and those that are different from each other. Students should understand that human beings have the same basic physical characteristics, the size, shape, and color of them is what makes us different from each other.

4. Begin a large group discussion by introducing the concept of a 'need', and how a need 'gets met'. Give examples such as:

   You are thirsty, this is a need. How does this need get met? (you take a drink of water)
   You are tired, this is a need. How does this need get met? (you go to sleep)
   You are restless, this is a need. How does this need get met? (you walk/run around)

OTM2160   ISBN: 9781487710224
© On The Mark Press

5. Divide students into pairs. Give them worksheets 4, 5, and 6. Students will begin by discussing with their partner, the questions on worksheet 4. They can complete the written component individually, followed up with a conversation piece at the bottom of worksheet 6, to be done in pairs. (The basic needs of a human being are food, water, air, and shelter.)

6. Give students worksheets 7 and 8. They will choose an animal that they have discussed with their partner in the previous activity, or one that they know a lot about. They will make a list of its basic needs and how its needs get met. On worksheet 8, students will compare how a human's basic needs get met to how their choice of animal gets its basic needs met. (A lesson on how to use a Venn diagram may be necessary).

7. Divide students into pairs. Give them worksheet 9. They will discuss, then record using pictures/words of ways humans are interfering with the ability of other living things to get what they need to live. (A large group discussion may be necessary prior to beginning this activity, about the meaning of "interfering"). *A follow up option is to come back together as a large group to discuss students' responses. Responses could be recorded on chart paper and posted in the room.*

Sample responses may be:
- humans are polluting waters that aquatic life, live in
- forests are being taken out for urban development, animals are losing their habitat

DIFFERENTIATION:

Slower learners may benefit by:

- Providing an oral response to the questions on worksheet 3, thus eliminating the written expectation.

- These learners could work in a small group with teacher direction to complete worksheet 7 and 8. One animal could be chosen by the group. Students' responses about its basic needs and how they are met could be recorded on chart paper. The Venn diagram to compare the ways a human and this animal gets its basic needs met could also be done on chart paper. This could be displayed in the classroom.

For enrichment, faster learners could design a representation of the environment that in which meets the needs of the animal they chose on worksheet 7. This could be done by making a diorama. Dioramas could be displayed in the classroom, or somewhere in the school for others to observe. Students could also provide a written component to accompany their dioramas, which would detail the animal and how its environment allows it to meet its basic needs.

Characteristics of the Human Body

Can you list some body parts that a human body has?
Can you explain what the body parts are used for? Let's
get listing!

Body Part	What is it used for?

OTM2160   ISBN: 9781487710224
© On The Mark Press

Look at Me, Look at You!

How are human bodies alike? How are they different from each other?

Partner Up!

A drawing of me:	A drawing of my partner:

OTM2160 ISBN: 9781487710224
© On The Mark Press

When I look at the pictures of myself and my partner, I see that we both have:

- _____

- _____

- _____

- _____

- _____

- _____

I look different from my partner in these 3 ways:

1. _____

_____

_____

_____

2. _____

_____

_____

_____

3. _____

_____

_____

_____

OTM2160   ISBN: 9781487710224
© On The Mark Press

Our Basic Needs

Think **Pair** **Share**

With a partner, do some thinking and sharing of ideas about the questions below.

"What are the four basic needs of a human being?"

"How does a person meet these needs?"

Record Your Thinking!

Basic Need #1: _____

Illustrate how this need is met:

Basic Need #2: _____

Illustrate how this need is met:

Basic Need #3: _____

Illustrate how this need is met:

OTM2160   ISBN: 9781487710224
© On The Mark Press

Basic Need #4: _____

Illustrate how this need is met:

Back in Conversation!

With your partner, talk about the basic needs of other living things. What do they need to live and grow? How do they meet their needs?

Needing a Comparison

Choose an animal that you know a lot about. What are its basic needs? How does this animal meet its needs?

Animal Name: _____

Basic Needs: _____

_____

_____

_____

_____

Illustrations of the animal meeting its basic needs:

OTM2160 ISBN: 9781487710224
© On The Mark Press

Use the Venn diagram to compare how you get your basic needs met to how your animal gets its basic needs met.

Remember the middle of the Venn diagram is used to list ways that are the same!

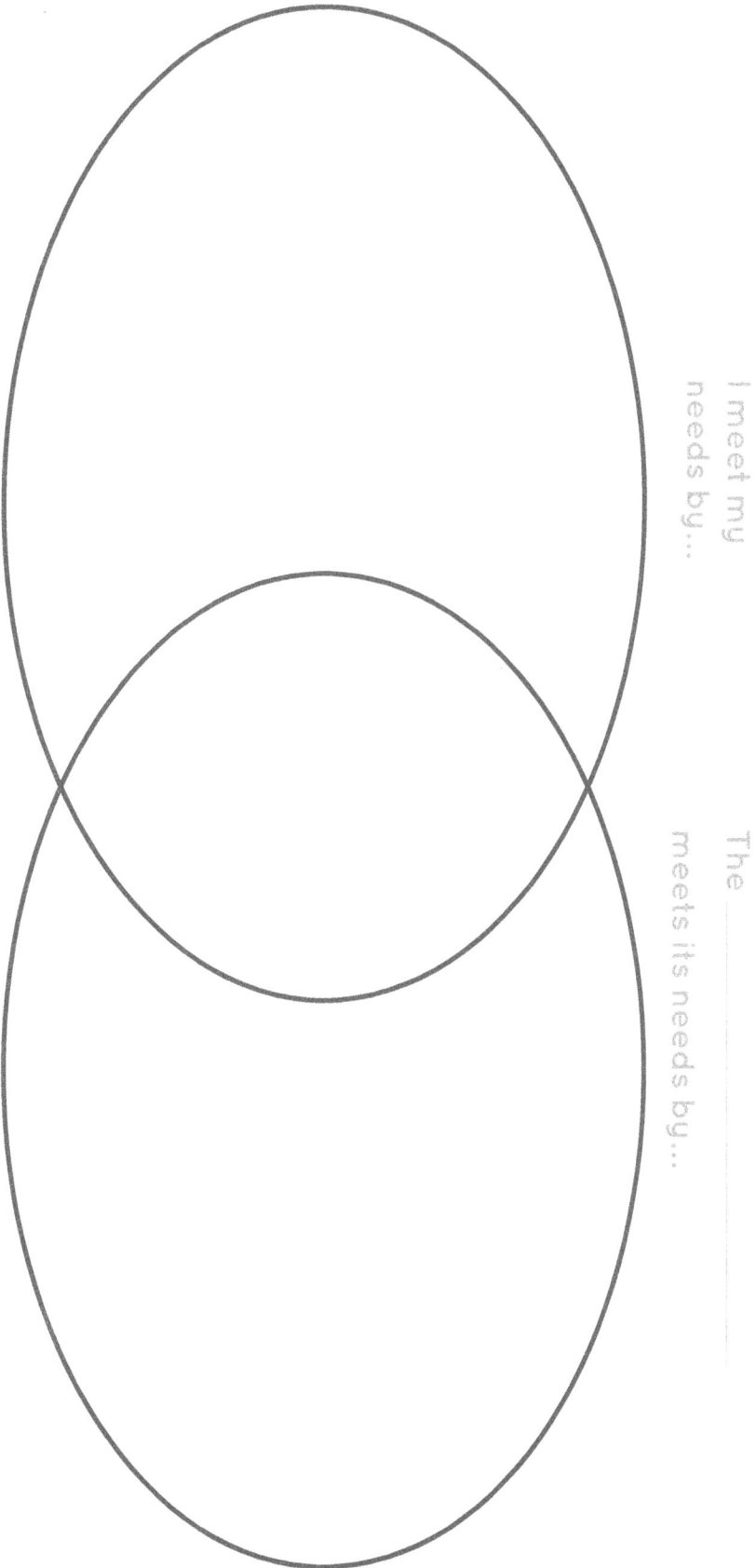

I meet my needs by...

The _____ meets its needs by...

Name:

Interfering with Nature

Think Pair Share

With a partner do some thinking and sharing of ways that humans might be interfering with other living things being able to meet their needs.

Record your ideas in the box below.

OTM2160 ISBN: 9781487710224
© On The Mark Press

NEEDS INTERTWINED!

LEARNING INTENTION:

Students will learn about what living things provide for each other, and the problems that could result from the loss of some kinds of living things.

SUCCESS CRITERIA:

- recognize that things used to meet needs are changed and returned in different forms
- identify what living things provided for other living things in order to have their needs met
- describe problems that could happen because of the loss of some kinds of living things

MATERIALS NEEDED:

- a copy of "Processing Needs" worksheet 1 and 2 for each student
- a copy of "Give and Take" worksheet 3, 4, and 5 for each student
- a copy of "A Missing Link" worksheet 6 for each student
- access to the internet, or local library
- chart paper, markers, pencils, pencil crayons

PROCEDURE:

*This lesson can be done as one long lesson, or can be divided into three shorter lessons.

1. Using worksheets 1 and 2, do a shared reading activity with the students. This will allow for reading practice and learning how to break down word parts in order to read the larger words in the text. Along with the content, discussion of certain vocabulary words would be of benefit for students to fully understand the passage.

Some interesting vocabulary words to focus on are:
- environment
- chlorophyll
- castings
- hydrated
- natural
- carbon dioxide
- nutrients
- liquid
- urine
- recyclers
- oxygen
- digest
- absorb
- returned
- breathes/breathing

2. Using worksheet 3, do a shared reading activity with the students. Along with the content, discussion of certain vocabulary words would be of benefit for students to fully understand the passage.

Some interesting vocabulary words to focus on are:
- exchange
- beaver dam
- insects
- nectar
- re-create
- gnaw
- stumps
- pollen
- provide

3. Give students worksheets 4 and 5. They can visit the local library or access the internet to gather information to help them answer the questions on worksheets 4 and 5. Upon completion, students can participate in a 'turn and talk' activity with a partner, where they share the information they have acquired on worksheet 5.

4. Divide students into pairs and give them worksheet 6. They will engage in a "think-pair-share" activity to discuss what the world would be like without certain living things in it. They will choose one of their ideas to illustrate the impact of its loss on the environment. Follow up with a large group discussion by asking some students to share their ideas. *Pose this question to the group: Who would be affected by the loss, and how?*

DIFFERENTIATION:

Slower learners may benefit by a reduction in expectations by the elimination of worksheet 5.

For enrichment, faster learners could make a list of jobs or hobbies that require knowledge about the needs of living things, and then choose one to illustrate and write about.

Processing Needs

You have learned a lot about the basic needs of living things. You know that all living things need water, food, air, and shelter to stay alive.

When a basic need is met, it gets used by a living thing, and it returns to the environment in a different form. Let's learn more about this idea!

Question:

What happens to the air that a human being uses to meet its need of breathing?

Answer:

The oxygen that a human breathes in is used to bring air into the body. When a human breathes out, the air comes out into the environment as carbon dioxide.

Question:

What happens to the air that a plant uses to meet its need of breathing?

Answer:

The carbon dioxide that a plant breathes in is used to bring air into it which is mixed with its chlorophyll to make plant food. When the plant breathes out, the air comes out into the environment as oxygen.

OTM2160 ISBN: 9781487710224
© On The Mark Press

Name:

Question:

What happens to vegetable scraps that an earth worm uses to meet its need of food?

Answer:

When an earthworm eats and then digests its food, its castings are left behind, which add nutrients to plants in the soil.

Question:

What happens to the liquid that wild animals use to meet their need for water?

Answer:

When wild animals drink water, they absorb what they need to be hydrated, and then it is returned to the Earth as urine.

DID YOU KNOW?

Red-bellied Woodpeckers meet their need of shelter by making nests inside dead trees. They chip away the wood to make a hole in the tree. Red-bellied Woodpeckers are natural recyclers in the environment because they use the left over woodchips by making a bed with them to lay their eggs on.

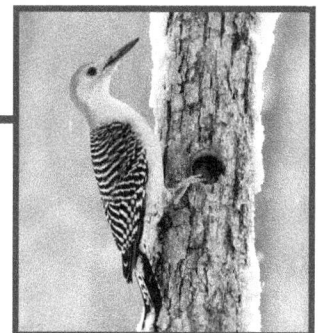

OTM2160   ISBN: 9781487710224
© On The Mark Press

Give and Take

While living things get their needs met, sometimes this action leaves behind something that another living thing could also use to meet its needs.

You have learned that as humans breathe out carbon dioxide, it is taken in by plants, which breathe out oxygen that humans take in. This is an exchange of needs.

DID YOU KNOW?

Beavers use their strong teeth to gnaw down trees which they need in order to make a beaver dam. The left over tree stumps can be useful to other living things. For example, a chipmunk will meet its need for shelter by making its home in a tree stump.

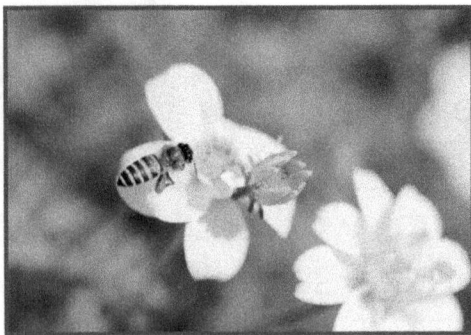

Insects and plants help each other meet their needs too. When insects land on plants, they carry and spread pollen that plants need to re-create. Plants provide shelter and food for some insects. For example, a bee gets the nectar it needs from a plant to make honey and the bee spreads the pollen of plants as it flies from flower to flower.

OTM2160 ISBN: 9781487710224
© On The Mark Press

Visit a library or use a computer to help you find out the answers to these questions about what some living things provide for each other.

Question:

What does a cow provide for other living things?

_____

_____

What needs does this meet?

_____

_____

_____

Question:

What does a maple tree provide for other living things?

_____

_____

What needs does this meet?

_____

_____

_____

It is your turn to ask some question of your own about what some living things provide for each other. Don't forget to find out the answers to your questions!

Question:

What does a _____ provide for other living things?

_____

_____

What needs does this meet?

_____

_____

Question:

What does a _____ provide for other living things?

_____

_____

What needs does this meet?

_____

_____

OTM2160   ISBN: 9781487710224
© On The Mark Press

A Missing Link

You have learned that living things have basic needs, and that sometimes these needs are provided by other living things.

With a partner, do some thinking and sharing of ideas about this question:

"How would the environment be different if there was a loss of some kinds of living things?"

Record Your Thinking!

Choose **one** living thing and illustrate all the ways the environment would be different without it.

OTM2160   ISBN: 9781487710224
© On The Mark Press

EXPLORE YOUR SENSES

LEARNING INTENTION:

Students will learn about the five senses and how they are used to identify objects in our world.

SUCCESS CRITERIA:

- identify the five senses and explain their function
- give examples of how each sense is used to recognize objects in our environment
- gather and record data in diagrams and in charts
- make connections to your environment

MATERIALS NEEDED:

- a copy of "The Five Senses" worksheet 1 for each student
- a copy of "Sensing the World Around Us" worksheet 2 and 3 for each student
- a copy of "What's the Sense?" worksheet 4 for each student
- a copy of "How Can You Sense It?' worksheet 5 and 6 for each student
- a copy of "How Does It Feel? worksheet 7 for each student
- a copy of "How Does It Taste? worksheet 8 for each student
- an apple, chart paper, markers, pencils
- chocolate pieces, pretzels, lemon wedges, rhubarb pieces (a piece for each student)

PROCEDURE:

This lesson can be done as one long lesson, or be divided into four shorter lessons.

1. Introduce to students our five different senses, ensuring their understanding/meaning of each one. Discuss how we use our senses to identify objects/the world around us. At this point, it would be beneficial for students to engage in a "knee to knee, eye to eye" activity where they turn and talk with a partner to brainstorm things that they can see, things that they can hear, things that they can smell, things that they can taste, and things that they can feel. Come back as a large group after each one to share responses orally. This will help to get ideas flowing. *An option is to record responses on chart paper.*

2. Give students worksheet 1, 2, 3 and 4 to complete.

3. Explain to students that we sometimes use more than one sense to recognize things. Giving them an example of an object such as an apple (pass it around), ask them what senses they could use to identify it. Give students worksheets 5 and 6 to complete.

4. Give students worksheet 7 to complete.

5. Discuss with students how the tongue has different areas on it that are sensitive to different tastes (sweet, bitter, bitter, and sour). An option is to give students a piece of a food to taste that will be detected by each area of their tongues. Have students determine which part of their tongue is sensing each food item. For example, give them a piece of chocolate, a pretzel, a lemon wedge, and a piece of rhubarb.

6. Give students worksheet 8 to complete.

DIFFERENTIATION:

Slower learners may benefit by listing only two or three things per sense on worksheets 2 and 3. An additional accommodation is to work in a small group with teacher support to read clues on worksheet 4.

For enrichment, faster learners could come together as a group to share their "own ideas" from worksheet 5. Using markers, they can display their ideas that they came up with on chart paper and then present it to the larger group. *A further enrichment activity would be to have these students create clues of their own and have a partner match it to one of the senses.*

OTM2160   ISBN: 9781487710224
© On The Mark Press

The Five Senses

Use the words in the word box to complete the sentences.

Word Box				
tongue	ears	nose	see	smell
hear	skin	eyes	taste	feel

1. We _____ with our _____.

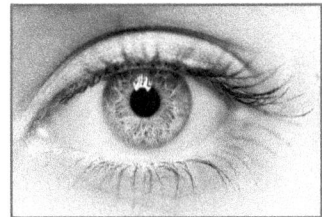

2. We _____ with our _____.

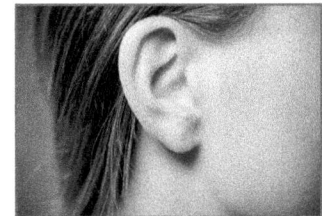

3. We _____ with our _____.

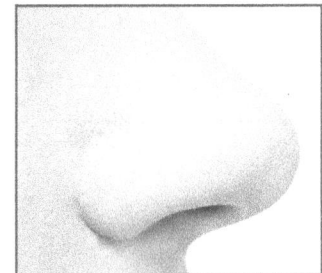

4. We _____ with our _____.

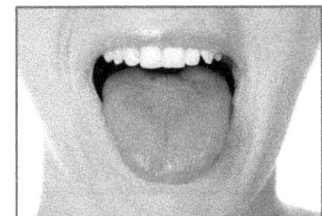

5. We _____ with our _____.

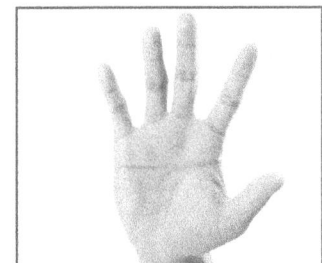

OTM2160   ISBN: 9781487710224
© On The Mark Press

Sensing the World Around Us

Our senses allow us to learn, to protect ourselves, and to enjoy our world. The five senses are **taste**, **touch**, **smell**, **sight**, and **hearing**.

Make a list of things that you can sense with your tongue, touch, nose, eyes, and ears.

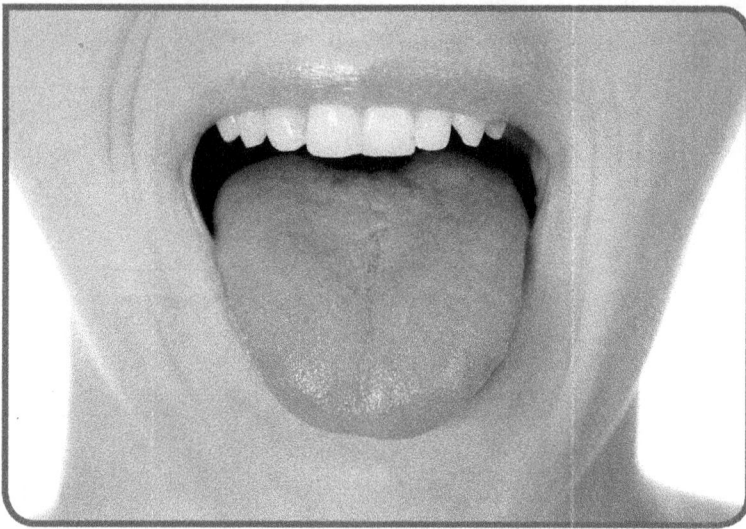

What can you taste?

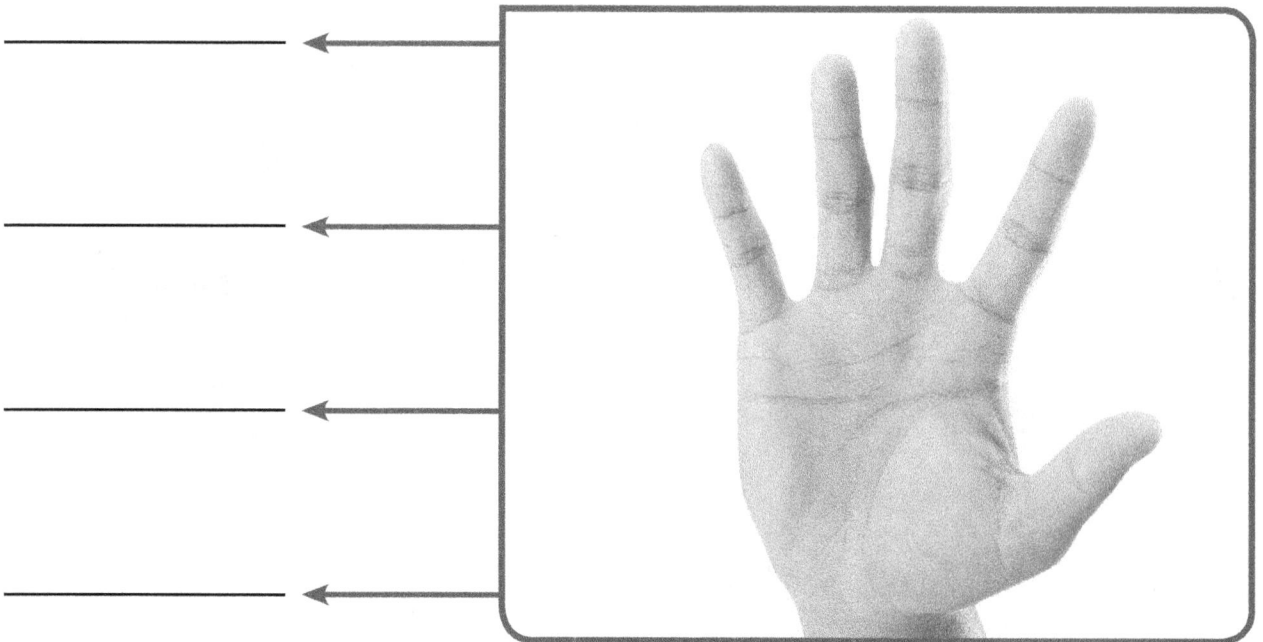

What can you feel?

OTM2160   ISBN: 9781487710224
© On The Mark Press

Name:

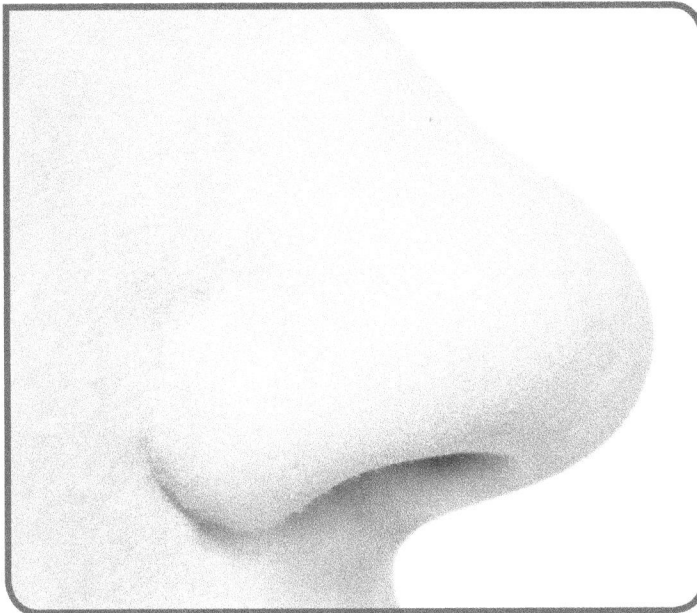

What can you smell?

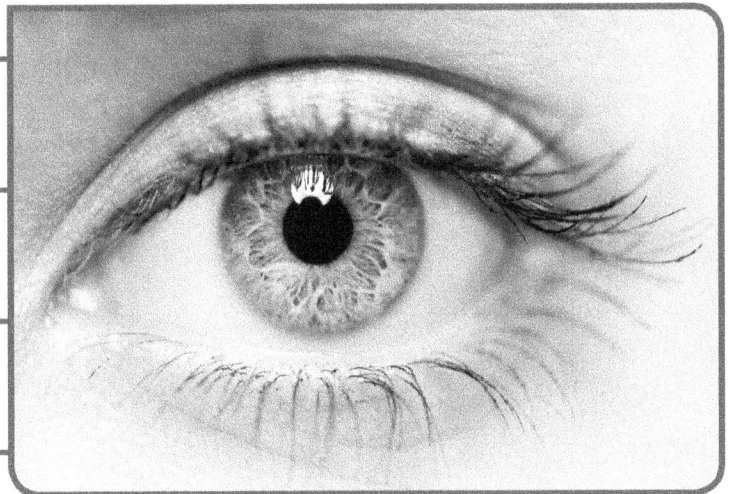

What can you see?

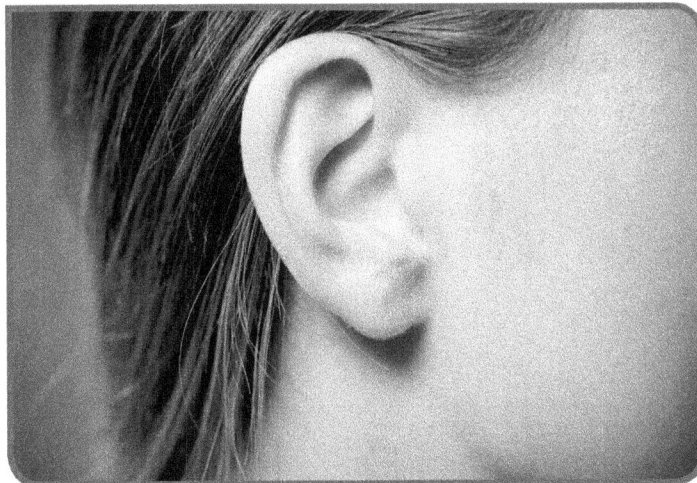

What can you hear?

OTM2160   ISBN: 9781487710224
© On The Mark Press

What's the Sense?

Fill in the blanks to the clues using the sense words from the Word Box.

Word Box

| see | hear | feel | smell | taste |

Clues:

I know that cotton candy is sweet because I can _____ it.

I know that birds chirp because I can _____ them.

I know that a rock is hard because I can _____ it.

I know that flowers are fragrant because I can _____ them.

I know that ice is cold because I can _____ it.

I know that thunder is loud because I can _____ it.

I know that the sun is bright because I can _____ it.

I know that a lemon is sour because I can _____ it.

OTM2160   ISBN: 9781487710224
© On The Mark Press

How Can You Sense It?

We can use more than one of our senses at a time to recognize things.

In the chart, list the senses that you **could** use to recognize what each object is. Then, add two of your own ideas.

OBJECT	SENSES USED
Popcorn	
Guitar	
Jello	
Rain	

You can use more than one sense to recognize these objects. But, can you sort them by using these clues?

1. Draw a **blue** line around three things that you can hear.

2. Draw a **red** line around three things that have a strong smell.

3. Draw a **green** line around two things that are a treat to eat.

4. Draw a **purple** line around one object that you cannot hear, smell, or taste.

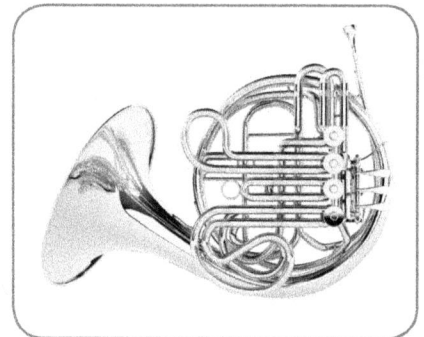

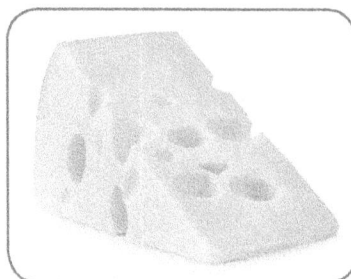

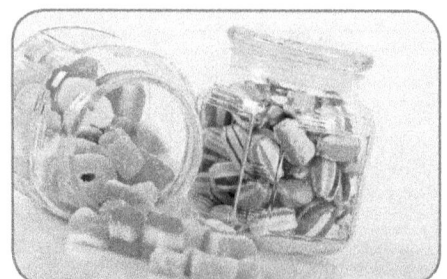

OTM2160   ISBN: 9781487710224
© On The Mark Press

How Does It Feel?

We can use any part of our skin to touch. By using our sense of touch, we can feel if something is hard or soft, wet or dry, or hot or cold.

Sort these objects by how they feel to the touch.

sponge	sandpaper	hot chocolate	snow	wood
ice cream	velvet	a cotton ball	a table	lemonade
fire	a brick	radiator	fur	a rock

Hard	Soft	Cold	Hot

Challenge!

Add some words of your own to fill up the table.

OTM2160   ISBN: 9781487710224
© On The Mark Press

Name:

How Does It Taste?

Did you know that different parts of your tongue are sensitive to different tastes? There are four taste zones on the tongue. They are sweet, **bitter**, salty, and **sour**.

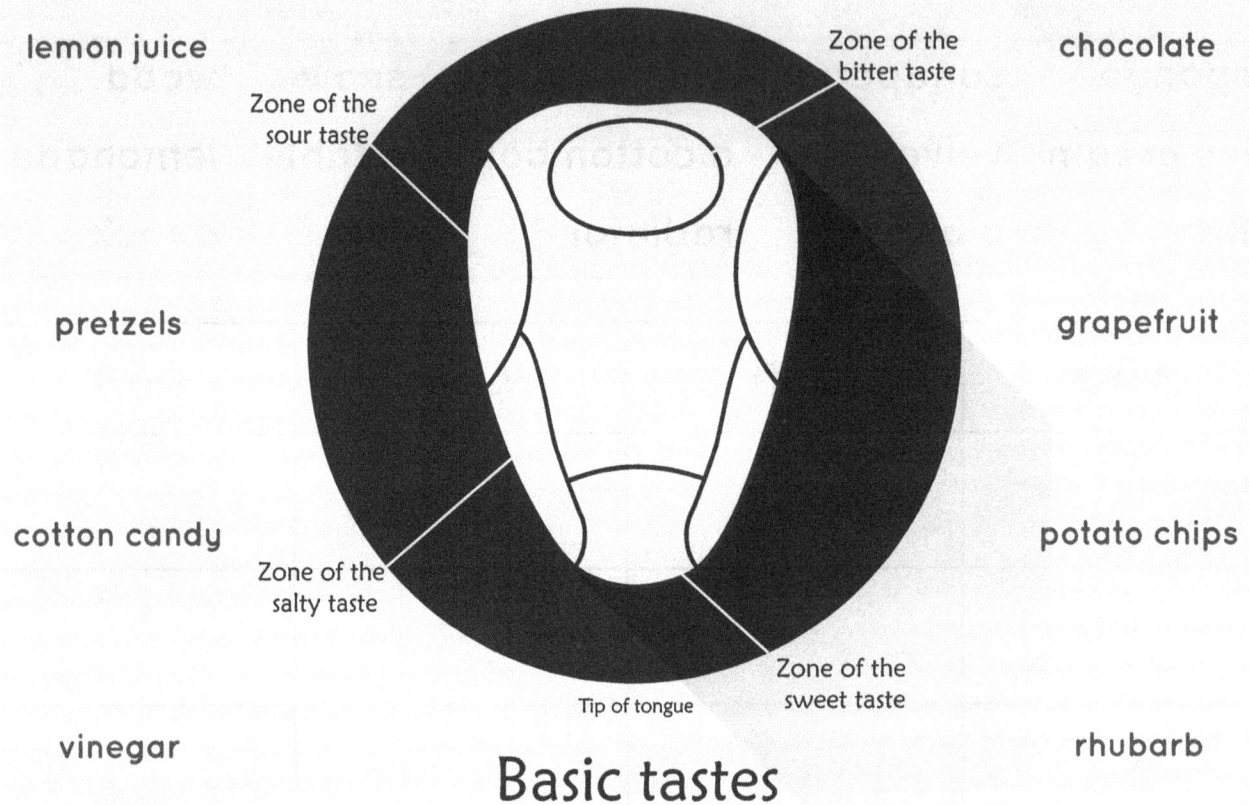

lemon juice

Zone of the
sour taste

Zone of the
bitter taste

chocolate

pretzels

grapefruit

cotton candy

Zone of the
salty taste

potato chips

Zone of the
sweet taste

vinegar

Tip of tongue

rhubarb

Basic tastes

Challenge!

1. Colour the sweet area orange. Colour the bitter area **blue**.

2. Colour the salty areas green. Colour the sour areas **purple**.

3. Draw a line from the food words to the area on the tongue that you would taste them.

OTM2160 ISBN: 9781487710224
© On The Mark Press

APPLYING YOUR SENSES

LEARNING INTENTION:

Students will learn how the senses are applied to describe and sort things in our environment.

SUCCESS CRITERIA:

- use the senses of sight, hearing, and smell to identify objects in a local area
- use the senses of touch, smell, taste, and hearing to recognize familiar objects
- memorize a group of objects that are within sight
- choose and use a sense to sort a group of objects
- use pictures and words to record responses
- make connections to the environment

MATERIALS NEEDED:

- a copy of "Getting a Sense of the Area" worksheet 1 and 2 for each student
- a copy of "Grab Bag of Sensible Fun" worksheet 3, 4, 5, and 6 for each student
- a copy of "How Sharp is Your Memory?' worksheet 7 for each student
- a copy of "Sort by Senses" worksheet 8 for each student
- 10 - 12 pairs of children's sunglasses (with masking tape covering the lenses)
- 6 timers, 6 large sheets of newspaper, chart paper, markers, clipboards, pencils

For touching activity:
- 6 grab bags of assorted objects (recognizable to the touch) such as a rock, paper clip, golf ball, a water bottle cap, eraser, a pine cone, a battery, etc.

For smelling activity:
- 4 - 6 plastic containers with tiny holes in lids, each containing coffee grinds
- 4 - 6 plastic containers with tiny holes in lids, each containing lemon wedges
- 4 - 6 plastic containers with tiny holes in lids, each containing onion slices
- 4 - 6 plastic containers with tiny holes in lids, each containing cinnamon

For tasting activity:
- 4 - 6 plastic containers with lids, each containing a few apple slices
- 4 - 6 plastic containers with lids, each containing a few onion slices
- 4 - 6 plastic containers with lids, each containing some shredded carrot
- 4 - 6 plastic containers with lids, each containing melon pieces

For listening activity:
- 4 - 6 plastic containers with lids, each containing coins
- 4 - 6 plastic containers with lids, each containing uncooked rice
- 4 - 6 plastic containers with lids, each containing small rocks
- 4 - 6 plastic containers with lids, each containing uncooked macaroni

For visual activity:
- 6 grab bags of 15 assorted small objects such as a rock, paper clip, golf ball, a water bottle cap, eraser, a pine cone, a battery, a pencil, a marble, a lollipop, a whistle, a spoon, popsicle stick, a juice box, an aromatic marker

PROCEDURE:

This lesson can be done as one long lesson, or done in seven or eight shorter lessons.

1. Give students worksheet 1, and a clipboard and pencil. Take them on a walk of the neighborhood. Instruct them to use their senses of sight, hearing, and smell to observe things in the area. An option upon returning to the classroom is to divide students into small groups to orally share their observations. Next, give students worksheet 2. Take them on a walk in a local park or forest area. Instruct them to use their senses of sight, hearing, and smell to

observe things in this area. Upon returning to the classroom, have students orally share their observations. Pose these questions to the large group:

Were any of the sights, sounds, smells the same in both the neighborhood and park/forest?

Were there any differences in the sights, sounds, smells in these areas?

2. Divide students into small groups of 4. Give each group a grab bag of assorted small objects and a pair of blinded sunglasses. Students will take turns to pull two objects out of the bag. Instruct them to use their sense of touch to describe it first, and then name it. They will draw and label their objects on worksheet 3.

3. Students will continue to work in groups of 4. Give each group 4 plastic containers (each with different contents) that have small holes punched through the lid to allow aroma out (use empty margarine containers and punch holes in the lids with a small nail and hammer). Students will take turns to smell the objects in the containers. Instruct them to use their sense of smell to describe it first, and then name the object inside. They will draw and label their objects on worksheet 4. *Materials suggested can be substituted for other familiar smelling things.

4. Students will continue to work in groups of 4. Give each group 2 pairs of sunglasses and 4 plastic containers (with lids, no holes), that have different food objects in them. Students will take turns (in pairs) to taste the objects in 2 of the containers. Instruct them to use their sense of taste to describe it first, and then name the object inside. They will draw and label their objects on worksheet 5. *Materials suggested can be substituted for other safe, familiar tasting foods. It is important to check before beginning this activity for any food allergies students may have. Also, it is important to explain to students that while you are encouraging them to use their sense of taste for this activity, it is not safe to do so outside

of this activity. It is important to always check first with a responsible adult.

5. Students will continue to work in groups of 4. Give each group 4 plastic containers (each with different non-food related contents). Students will take turns to listen to the sound the objects in the containers make as they are shaken or moved around. Instruct them to use their sense of hearing to describe it first, and then name the object inside. They will draw and label their objects on worksheet 6. *Materials suggested can be substituted for other familiar sounding things.

6. Students will continue to work in small groups. Give each group a bag of 15 assorted small items, a timer, and a clipboard and pencil. They will empty the contents of the bag and give themselves 2 minutes to memorize what they see. Once the time is up, the objects are to be covered while students record from memory the objects that they saw, using worksheet 7. *Materials suggested can be substituted for other familiar objects.

7. Give students worksheet 8. Using the objects from the grab bag in the previous activity, students will sort them according to one or more of the senses. A follow-up option is to have students share their sorting rule with a classmate to compare different ways to sort.

DIFFERENTIATION:

Slower learners may benefit by working as a small group and having a grab bag of only 10 assorted objects. An additional accommodation would be to work as a small group with teacher support to complete worksheet 8.

For enrichment, faster learners could make a list of visual clues that describe a classmate. These lists could be read to the large group, who will make a guess of who the clues are about.

OTM2160 ISBN: 9781487710224
© On The Mark Press

Name:

Getting a Sense of the Area!

Take a walk around your neighborhood. Get a sense of the area by telling about what you see, hear, and smell.

I see...

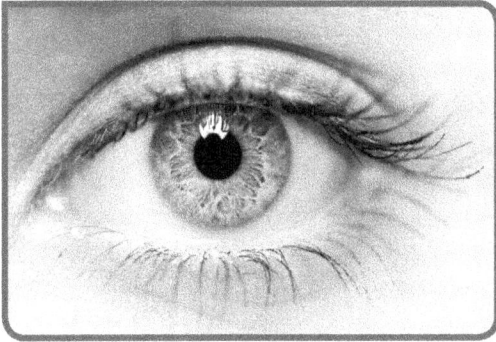

I hear...

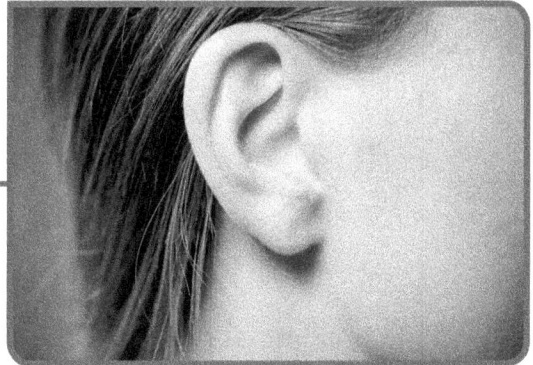

I smell...

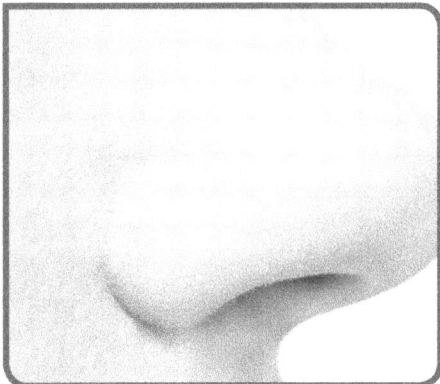

OTM2160   ISBN: 9781487710224
© On The Mark Press

Now take a walk around a local park or forest. Get a sense of this area by telling about what you see, hear, and smell.

In the _____:

I see...

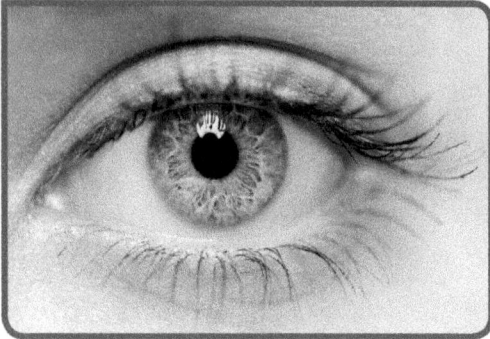

I hear...

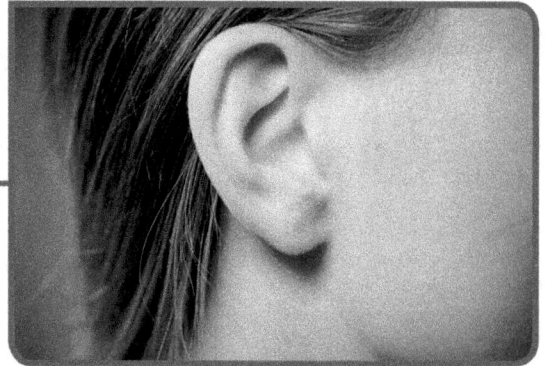

I smell...

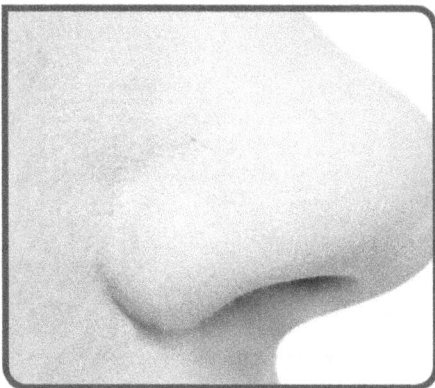

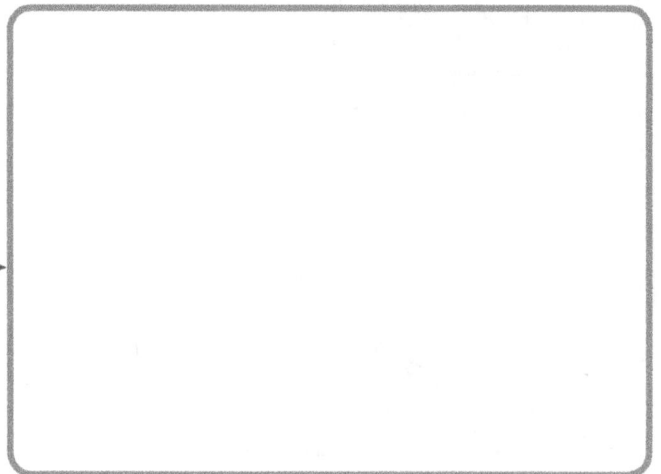

OTM2160   ISBN: 9781487710224
© On The Mark Press

Grab Bag of Sensible Fun!

Activity #1 – Touch and Guess

1. Put the blinded sunglasses on.

2. Reach into the grab bag and pull out one object.

3. Use your sense of touch to **describe** the object.

4. Guess what the object is.

5. Take off the sunglasses and check your guess. Draw and label the object in the box below.

6. Repeat steps 2, 3, 4, and 5.

7. Pass the grab bag and sunglasses to another group member.

Object #1	Object #2

Activity #2 – Smell and Guess

1. Use your sense of smell to **describe** the object inside container #1.

2. Guess what the object is. Pass the container to another group member.

3. After each group member has described and made a guess about what the object is, open the lid of the container to check your guess.

4. What is the object? Draw it and label it in the box below.

5. Repeat all the steps for each container on the table.

Object #1	Object #2

Object #3	Object #4

OTM2160   ISBN: 9781487710224
© On The Mark Press

Activity #3 – Taste and Guess

1. Two group members will put blinded sunglasses on. Another group member will open the lid of a tasting container.

2. The two group members will plug their noses and reach into the container to take out an object of food.

3. Use your sense of taste to **describe** the food.

4. Guess what the food is.

5. Take off the sunglasses and check your guess. Draw and label the food object in the box below.

6. Repeat all the steps using another tasting container.

7. Switch roles with the other two group members so they can become taste testers!

Object #1	Object #2

Activity #4 – Listen and Guess

1. Use your sense of hearing to **describe** the object inside container #1.

2. Guess what the object is. Pass the container to another group member.

3. After each group member has described and made a guess about what the object is, open the lid of the container to check your guess.

4. What is the object? Draw it and label it in the box below.

5. Repeat all the steps for each container on the table.

Object #1 _____	**Object #2** _____
Object #3 _____	**Object #4** _____

OTM2160 ISBN: 9781487710224

Name:

How Sharp is Your Memory?

It is time to play a memory game!

You'll need:

- a grab bag of 15 objects
- a table
- a pencil
- a large sheet of newspaper
- a timer

What to do:

1. Empty out the grab bag of objects onto the table.

2. Set the timer for 2 minutes. Try to remember the objects that you see.

3. When the time is up, cover the objects with the sheet of newspaper.

4. In the box below, record the objects that you saw.

Sort by Senses

Use the grab bag of objects from worksheet 7. Show how you can sort them. What sense will you use to sort?

To sort the objects, I used the sense of _____.

OTM2160   ISBN: 9781487710224
© On The Mark Press

ANIMAL SENSES

Learning Intention:

Students will learn about how animals use their senses to survive in the natural world.

Success Criteria:

- recognize that animals need to use their senses to survive in nature
- research and describe how an animal uses its senses
- orally communicate research findings with other classmates
- listen to and record interesting facts about how animals use their senses to survive

Materials Needed:

- a copy of "Sensing to Survive!" worksheet 1 and 2 for each student
- a copy of "Making Use of Its Senses" worksheet 3 and 4 for each student
- a copy of "Sensing the Animal Facts!" worksheet 5 for each student
- access to the internet, or local library
- chart paper, markers, pencils, clipboards

Procedure:

*This lesson can be done as one long lesson, or done in four or five shorter lessons.

1. Using worksheets 1 and 2, do a shared reading activity with the students. This will allow for reading practise and learning how to break down word parts in order to read the larger words in the text. Along with the content, discussion of certain vocabulary words, would be of benefit for students to fully understand the passage.

   Some interesting vocabulary words to focus on are:

   - prey
   - sensors
   - echo
   - perched
   - chemicals
   - texture
   - scan
   - chemoreceptors
   - tongue
   - tentacles
   - predators
   - nectar
   - echolocation

2. Give students worksheets 3 and 4, and a clipboard and pencil. They can visit the local library or access the internet to gather information to help them find out how their chosen animal uses its senses to help it survive in nature (this animal was researched while doing Describing an Animal, worksheets 5 and 6 (p. 21, 22), from In the Animal Kingdom lesson plan. *An option is to choose a different animal from the one they previously researched.

3. Divide students into groups of 5, and give them worksheet 5. Each student in the group will orally present their research on how their chosen animal uses its senses to survive in nature. Instruct students to record **one** interesting fact as **each** group member presents their information. Come back as a large group to share some of the facts students heard about, in order to promote some rich discussion and the usage of scientific content words.

4. As an activity to enhance the learning about animal senses, show students Wild Kratts episode "Platypus Café". Episode can be accessed at www.youtube.com

   *In this episode, the platypus uses electrosense to detect the location of objects. The rattle snake uses heat sense to see objects more clearly.

Differentiation:

Slower learners may benefit by working as a small group with teacher support and direction to complete worksheets 3 and 4, using one animal. Sections of the assignment could be divided between these students to lessen the work load.

For enrichment, faster learners could illustrate how the platypus and the rattlesnake each uses their "extra sense" to survive in nature. Encourage students to add a descriptive sentence.

Sensing to Survive!

Did you know that animals have senses too? They use their senses to help them survive in nature. Let's learn more about this!

Fast Facts!

Did you know that a snake uses its tongue to "taste" the air? When a snake flicks out its tongue, it is able to sense smells in the air. When the snake's tongue goes back into its mouth, it touches the top of its mouth and tells the snake what it smells.

A snake uses its tongue to smell prey and predators that are close by.

Bats can tell the size, shape, and texture of an insect all from its echo.

Have you ever heard of **echolocation**? Echolocation is what bats use to find their way in the dark and to find food. Bats send out sound waves with their mouth or nose. When a sound wave hits an object, an echo comes back. This helps the bat to know what the object is, where it is, and how close it is.

OTM2160   ISBN: 9781487710224
© On The Mark Press

The star nose mole is an animal that has a great sense of touch. This is because it has 22 tentacles that form a star shape on its nose. The star nose mole cannot see well. It uses its tentacles to feel around and to find its prey like worms, small fish, and insects.

A star nose mole can touch as many as 12 different objects per second!

Did you know that a butterfly tastes with its feet? A butterfly's feet have sensors that can taste the nectar in a plant. If the nectar tastes good, the butterfly will eat it. A female butterfly tastes a plant to find out if it is a good place to lay eggs. She uses the chemoreceptors on her legs to find the right match of plant chemicals. When she finds the right plant, she lays her eggs.

Birds of prey such as eagles and hawks have very good eyesight. They can scan the Earth as they are flying or perched, and spot their prey from far away. Did you know that an eagle can turn its head almost all the way around, and that its eyes turn too?

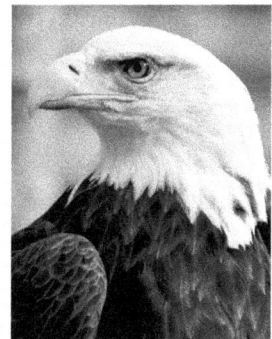

Making Use of Its Senses

Take a look back at the work you did on **Describing an Animal**, worksheets 5 and 6.

Visit a library or use a computer to find out how this animal uses its senses to survive in nature.

The _____

uses its **sense of sight** to _____

_____

_____

_____

_____

_____

The _____

uses its **sense of hearing** to _____

_____

_____

_____

_____

_____

OTM2160　ISBN: 9781487710224

Name:

The _____
uses its **sense of smell** to _____

_____

_____

_____

The _____
uses its **sense of taste** to _____

_____

_____

_____

The _____
uses its **sense of touch** to _____

_____

_____

_____

OTM2160   ISBN: 9781487710224
© On The Mark Press

Sensing the Animal Facts!

Getting the Facts!

Listen to your group members as they tell about how their animal uses its senses to survive in nature. Record some facts that you learned about.

1. _____

_____

_____

2. _____

_____

_____

_____

3. _____

_____

_____

_____

4. _____

_____

_____

OTM2160   ISBN: 9781487710224
© On The Mark Press

PROTECTING THE SENSES

LEARNING INTENTION:

Students will learn about the protective parts of the eye and ear; and about the importance of protecting our senses that contribute to our safety in daily living activities.

SUCCESS CRITERIA:

- identify and label some parts of the eye, and determine their use in protecting the eye
- identify some parts of the ear and determine their use in protecting the ear
- describe the hearing ability of an animal
- recognize some parts of the nose and their purpose
- recognize ways that we can protect our senses of taste, smell, and touch
- describe the affects of losing a sense has on daily living
- recognize ways that the five senses contribute to our daily lives

MATERIALS NEEDED:

- a copy of "The Eye" worksheet 1 and 2 for each student
- a copy of "The Ear" worksheet 3 and 4 for each student
- a copy of "Honing in on Hearing" worksheet 5 and 6 for each student
- a copy of "Being Nosey About the Nose" worksheet 7 and 8 for each student
- a copy of "Protecting the Others" worksheet 9 for each student
- a copy of "Sensible Safety!" worksheet 10 for each student
- about 10-12 small hand mirrors
- access to music

- access to the internet, or local library
- access to television/ video machine and video
- 10-12 pairs of children's sunglasses (with masking tape covering the lenses)
- different grab bags of small assorted items (one for each student)
- chart paper, markers, pencils, pencil crayons, clipboards

PROCEDURE:

*This lesson can be done as one long lesson, or done in eight or nine shorter lessons.

1. Give students worksheet 1 and a small hand mirror. They will take a close look at their own eye. They will draw a diagram of their eye, labeling its basic parts using words from the Word Box.

2. Using worksheet 2, do a shared reading activity with the students. This will allow for reading practise and learning how to break down word parts in order to read the larger words in the text. Along with the content, discussion of certain vocabulary words would be of benefit for students to fully understand the passage.

   Some interesting vocabulary words to focus on are:
   - eyebrows   • eyelids   • eyelashes   • pupil
   - iris   • protect   • moist   • sweeping
   - center   • muscle   • sweat   • blinking
   - harm

3. Discuss with students the importance of protecting their eyes and ways that they can actively do that (e.g., wear sunglasses, wear corrective lenses if needed, read with the right amount of light, get regular eye exams, wear protective sporting eye gear).

4. Give students worksheet 3. As they listen to music, they will experiment by covering/cupping the ear in different ways. Students can discuss the effects of the size and shape of the ear. (The shape of the outer ear helps to funnel sound to the inner parts of the ear, then nerve impulses send messages to the brain to interpret the sound.

5. Using worksheet 4, do a shared reading activity with the students. This will allow for reading practise and learning how to break down word parts of larger words in the text. Along with the content, discussion of certain vocabulary would benefit students.

   Some interesting vocabulary words to focus on are:
   - outer ear
   - sound vibrations
   - inner ear
   - collects
   - curled
   - protected
   - ear canal
   - nerve impulses
   - ear drum
   - cochlea
   - separates
   - middle ear

6. Discuss with students the importance of protecting their ears and ways that they can actively do that (e.g. wear ear plugs when in areas with excessive noise, listen to music or television at an appropriate volume, clean ears regularly, get medical check-ups).

7. Give students worksheet 5. They will circle the animals that have a good sense of hearing, and explain their choices. (All animals should be circled except the hippopotamus and the turtle.)

8. Give students worksheet 6. They can visit the local library or access the internet to gather information on the hearing ability of their chosen animal.

9. Using worksheets 7 and 8, do a shared reading activity with the students. This will allow for reading practise and learning how to break down word parts of larger words in the text. Along with the content, discussion of certain vocabulary would benefit students.

Some interesting vocabulary words to focus on are:
- nostrils
- organ
- throat
- mucous membrane
- lungs
- sinuses
- blood vessels
- moistens
- trachea
- nasal cavity
- healthy
- septum
- mucous
- breathe

10. Divide students into pairs and give them worksheet 9. They will engage in a 'think-pair-share' activity to discuss and record things we can do to protect our senses of touch, taste, and smell. *Inform students that we touch with all parts of our body (via our skin). An option is to come back as a large group to share their responses.

11. To teach students what it would be like to not be able to see or hear, but still have to function by use of other senses, try these activities:
    - have students put the blinded sunglasses on and sort a bag of items (having to determine their own sorting rule)
    - have students watch a television show/video without the volume on, afterwards have them try to discuss the events, then watch it with the volume on to see how accurate they were

12. Engage students in a discussion about ways that people adapt to the loss of a particular sense. Also discuss if there are any limitations to our senses. (They should be able to draw upon their own experiences from the previous activities.)

13. Give students worksheet 10 to complete.

DIFFERENTIATION:

Slower learners may benefit by working with a strong peer in order to 'navigate' the research required to complete worksheet 6.

For enrichment, faster learners could create a poster that details the importance of protecting our senses.

OTM2160 ISBN: 9781487710224
© On The Mark Press

The Eye

Take a Look!

Use a small mirror to take a close look at your eye.

1. What parts do you see?

2. Draw your eye.

3. Use the words in the Word Box below to label some of its parts.

---

eyebrow   eyelid   eyelashes   pupil   iris

OTM2160   ISBN: 9781487710224
© On The Mark Press

Have you ever wondered what these parts of your eye do? They protect the eye. Let's learn more about this!

Eyebrows protect our eyes from water or sweat that could drip down and come into our eyes.

Eyelashes protect our eyes by sweeping away dust and sand that can get into the eyes and harm them. Eyelashes also help to keep sweat or rain out of the eyes.

Eyelids protect our eyes by opening and closing over them to keep them moist. This is called blinking.

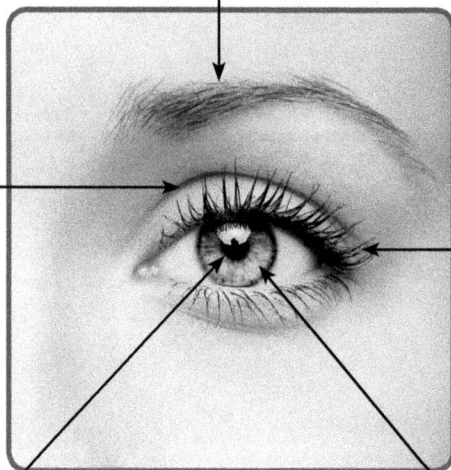

The **pupil** is the center of the eye.

When there is a lot of light, the pupil gets smaller to protect the eye from getting too much light in it.

When there is not a lot of light, the pupil gets bigger so that it can let in as much light as possible to give the eye more seeing power.

The **iris** is the colour part of the eye.

The iris is like a muscle for the eye. It helps the pupil to grow bigger, or get smaller.

It works with the pupil to protect the eye from getting too much light in it, and it works to let more light in if the eye needs more seeing power.

80

OTM2160 ISBN: 9781487710224
© On The Mark Press

The Ear

Listening In!

Take a close look at the ear. Why do you think it is shaped the way it is?

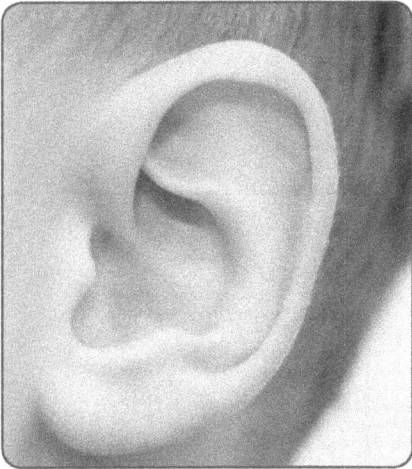

I think the ear is shaped this way because _____

_____

_____

_____

Try this!

Your teacher will play some music.

1. Cup your hand in front of your ear. Listen to the music.

2. Cup your hand behind your ear. Listen to the music.

3. With a classmate, talk about what happened.

With your partner, talk about the shape and size of the ear. Why are ears shaped like a cup? Does the size of the outer ear make a difference?

Our ears have three main parts. They are the outer ear, the middle ear, and the inner ear. Each part does something to protect the ear. Let's learn more about this!

The **outer ear** is the part that you can see. It opens into the ear canal. The outer ear collects sounds and it is like a path for the sound to get to the inside of the ear. The **ear canal** is protected by tiny hairs. It makes wax that traps dirt and dust from getting into the ear.

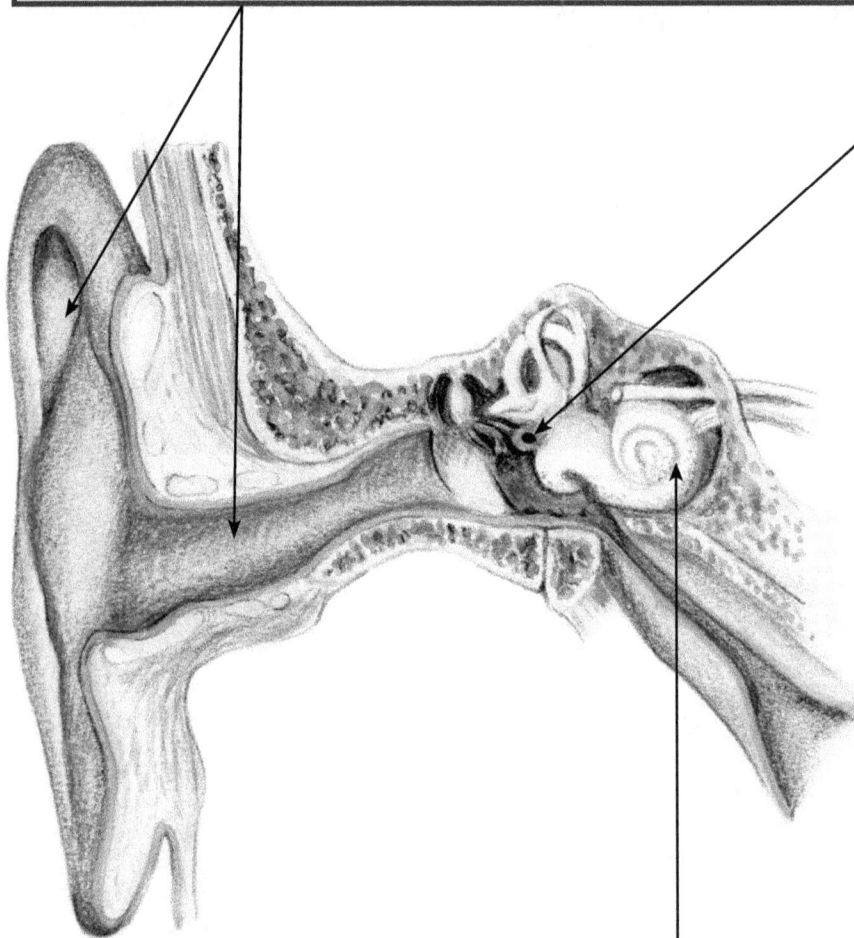

The **ear drum** separates the outer ear from the middle ear. The **middle ear** has three small bones in it that move to send sound vibrations to the inner ear.

The **inner ear** has a small curled tube in it called the **cochlea**. The cochlea has tiny hairs that move. They turn the vibrations sent by the middle ear into nerve impulses that go to the brain. The brain tells you what you are hearing.

OTM2160 ISBN: 9781487710224

Honing in on Hearing

Look at the ears of the animals in the pictures. Circle the ones that you think have the best hearing.

Why do you think the animals you circled have a good sense of hearing?

_____

_____

_____

Choose one of the animals you circled on Worksheet 5.
Visit a library or use a computer to find out more about
this animal's sense of hearing.

The _____

I learned that _____

_____

_____

_____

I also learned that _____

_____

_____

_____

Draw a picture to go with one of the facts you learned
about this animal.

OTM2160   ISBN: 9781487710224
© On The Mark Press

Name:

Being Nosey About the Nose

Your nose is the organ that lets you smell, and it helps you taste because it smells the food you are eating. The nose has another job too. It breathes in air to your lungs that are inside your body. Let's learn more about the parts of your nose that help you to do these things!

The **nostrils** are the openings to your nose. They breathe in air that travels to your **nasal cavity**. From there, air goes down your throat into the **trachea**, which is like a windpipe going to your lungs. Then air fills your lungs so your body can breathe.

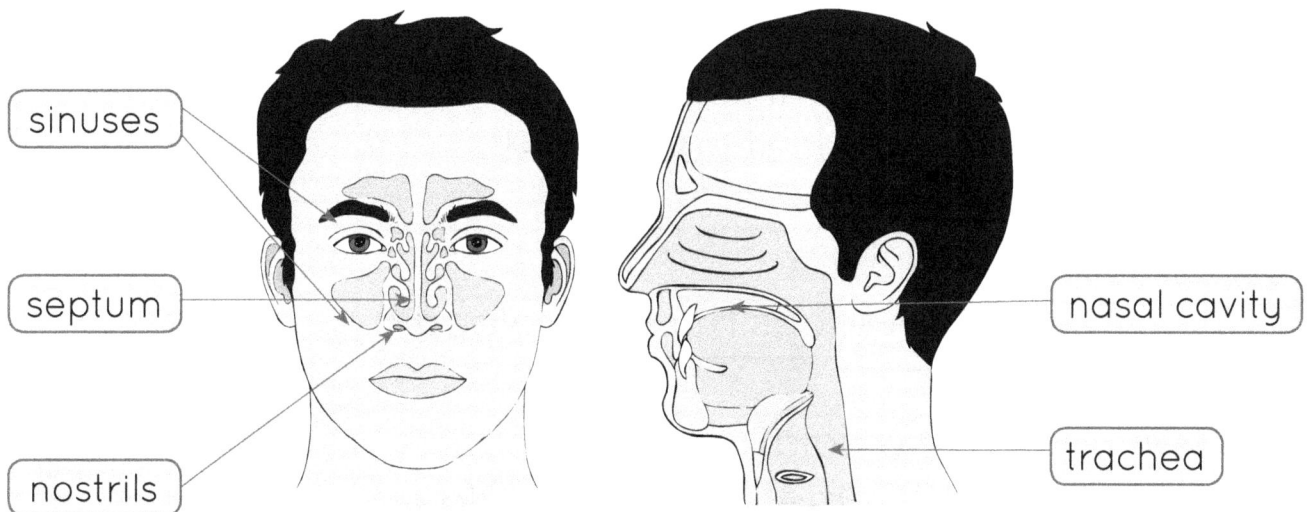

sinuses

septum

nostrils

nasal cavity

trachea

DID YOU KNOW?

The inside of your nose has mucous membrane. When you breathe in, the membrane warms and moistens the air. The membrane and the tiny hairs inside your nose trap dust and dirt that would not be healthy for your lungs. Sinuses in your nose store mucous.

OTM2160   ISBN: 9781487710224
© On The Mark Press

Question:
Why does my nose run?

Answer:

If you have a cold, your nose makes extra mucous to keep the germs from getting into your lungs. It could make so much that it runs out of your nose!

And did you know that if you are outside on a cold day, your nose warms up the cold air before it gets to your lungs? The blood vessels in your nostrils get bigger to warm the air, but this extra blood flow makes more mucous that runs out of your nose!

Question:
Why does my nose bleed?

Answer:

When you have a cold, you sneeze and blow your nose. Sometimes this could make the tiny blood vessel in your nose break and your nose might bleed.

And did you know that when the indoor air is very dry, the inside of your nose gets crusted and itchy. When you scratch that itchy nose, it may start to bleed.

An injury to the nose could also make it start to bleed. So take care and wear protective gear when playing sports!

OTM2160   ISBN: 9781487710224
© On The Mark Press

Name:

Protecting the Others

Think **Pair** **Share**

With a partner, so some thinking and sharing of ideas about the ways we can protect our senses of touch, taste, and smell. Record your ideas.

TASTE

SMELL

TOUCH

Sensible Safety!

You have learned a lot about the five senses, and how important it is to protect them.

Now, illustrate ways that our five senses keep us safe and help us in our daily lives.

OTM2160 ISBN: 9781487710224
© On The Mark Press

THE HUMAN BODY

LEARNING INTENTION:

Students will learn about the location and function of major parts in the human body and how they are used to meet our needs and explore our world.

SUCCESS CRITERIA:

- identify and label parts of the outer human body
- research the location and function of the five major organs inside the human body
- determine the location and function of some outer and inner body parts
- analyze how our body parts help us to meet our needs and explore our world

MATERIALS NEEDED:

- a copy of "Body Basics" worksheet 1 and 2 for each student
- a copy of "Building on the Basics" worksheet 3 for each student
- a copy of "A Look Inside" worksheet 4 and 5 for each student
- a copy of "Hard at Work" worksheet 6 and 7 for each student
- access to the internet, or local library
- scissors, glue, clipboards, pencils, timers

PROCEDURE:

This lesson can be done as one long lesson, or be divided into four shorter lessons.

1. Give students worksheets 1 and 2 to complete. Upon completion, engage students in a discussion about the importance of protecting the head (it is where most of our senses are located). Pose these questions: When should we protect our head? How should we do it?

2. Play a game of 'Simon Says' in order to get students thinking and learning about their body parts (use specific body parts such as wrist, thumb, ankle, etc.). Give students worksheet 3 to complete.

3. Give students worksheets 4 and 5. They can visit the local library or access the internet to gather information on the body's major internal organs (heart, brain, lungs, liver, and kidneys). *An option is to have a discussion about the functions of these internal organs before giving them worksheet 5 to complete.*

4. Give students worksheet 6 and 7. They will determine the location and function of some parts of the human body. As a follow up discussion, pose this question to the students: How do our body parts help us to explore and learn more about our environment/ world?

DIFFERENTIATION:

Slower learners may benefit by working as a small group with teacher support and direction to complete worksheet 3. To lessen the work load, each student in the group could be assigned an organ to research, and then share the findings with the small group in order to complete the worksheet. An additional accommodation would be for these learners to continue working in a small group with teacher assistance in order to read the clues on worksheet 4.

For enrichment, faster learners could take their pulse to determine how quickly their heart is beating after doing different activities for two minutes each (with a rest period in between to bring their heart rate back to normal). Suggested activities are walking, going up and down stairs, running on the spot, skipping, dancing to music. Pose this question: Which activity gets your heart rate up the most?

OTM2160   ISBN: 9781487710224
© On The Mark Press

Body Basics

Cut out the body labels below. Glue them on the body in the correct place.

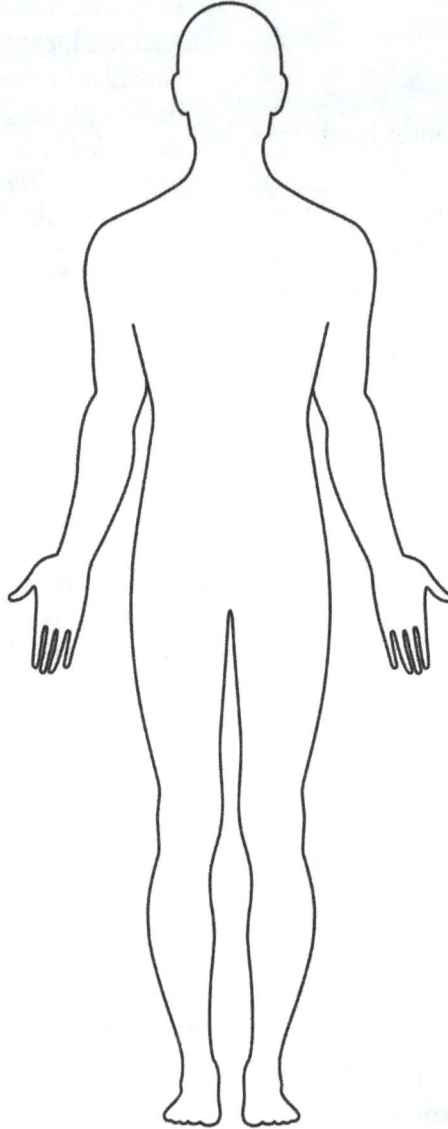

leg	fingers	neck	foot
head	chest	hand	arm

OTM2160   ISBN: 9781487710224

Name:

1. In what part of the body are your eyes?

   My eyes are in my _____. They are

   used for _____.

2. In what part of the body are your ears?

   My ears are on my _____. They are

   used for _____.

3. In what part of the body is your nose?

   My nose is on my _____. It is

   used for _____.

4. In what part of the body is your tongue?

   My tongue is in my _____. It is

   used for _____.

5. Where is your skin?

   My skin is _____. It is

   used for _____.

OTM2160   ISBN: 9781487710224
© On The Mark Press

Name:

Building on the Basics

Your outer body has many parts to it. Use the words in the word box to help you label parts on the human body.

shoulder	waist	leg	ears	ankle
elbow	mouth	nose	knee	fingers
hair	arm	eyes	foot	throat

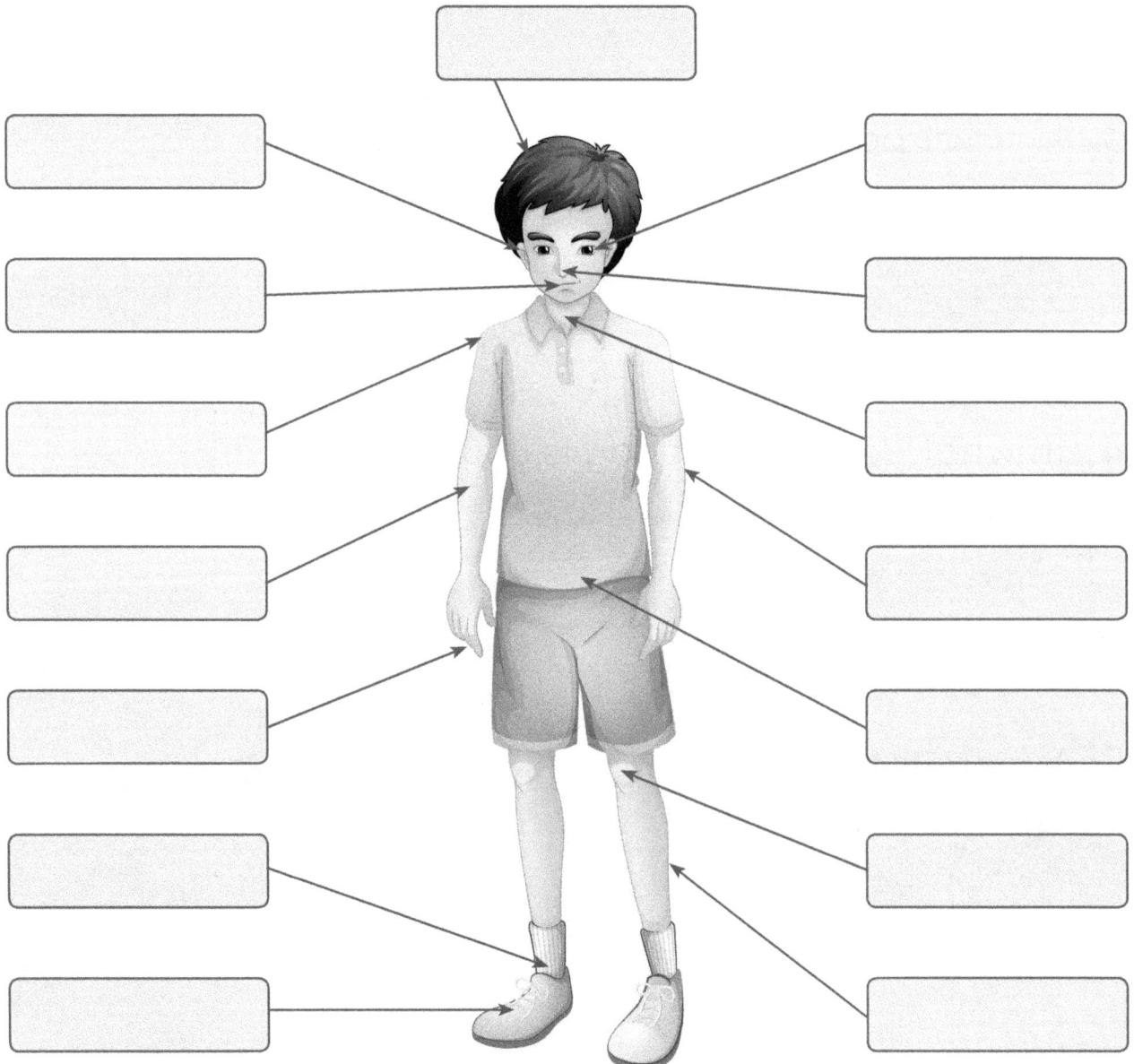

OTM2160   ISBN: 9781487710224
© On The Mark Press

A Look Inside

The **brain**, the **heart**, the **lungs**, the **kidneys**, and the **liver** are organs inside the human body. Visit a library or use a computer to find out more about them and where they are in your body.

Draw a line from each organ to show where it is inside the body.

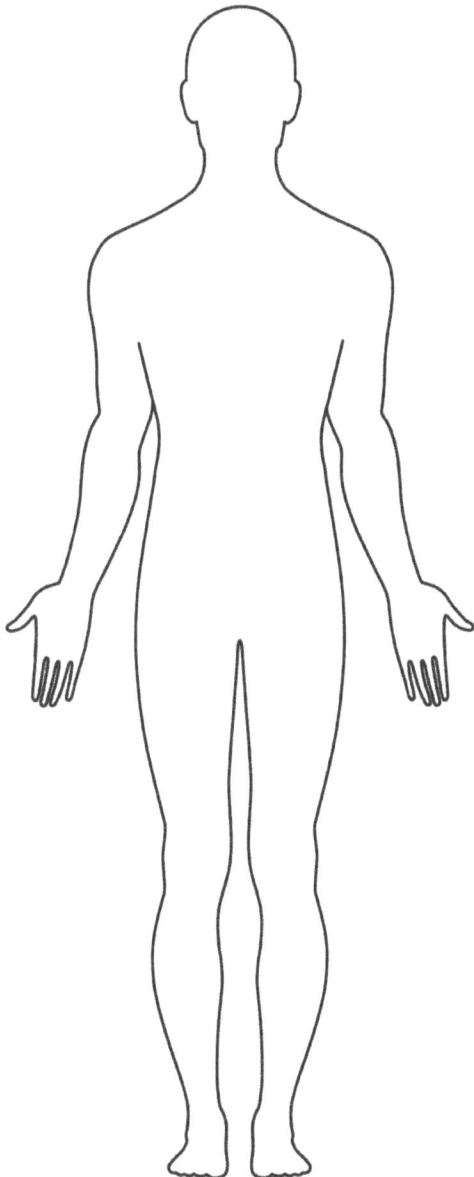

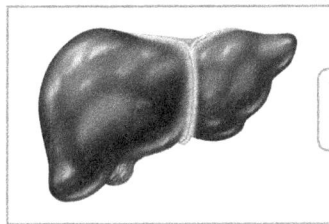 liver

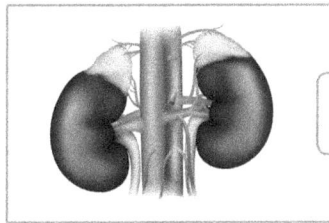 kidneys

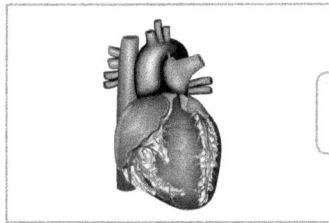 heart

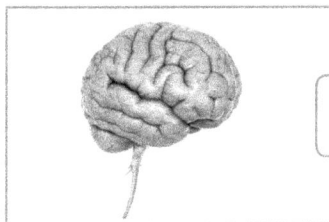 brain

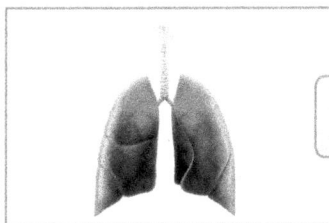 lungs

Name:

Draw a line from each organ to match what it does inside your body.

This organ is the nerve center in your body. It tells you what to think and how to act.

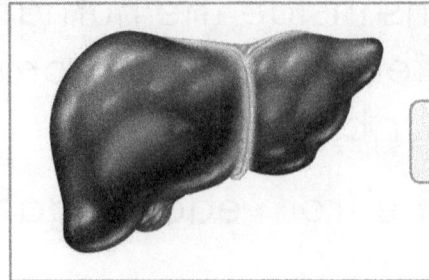

liver

This organ is like a pump inside your body. It works hard when you are moving fast.

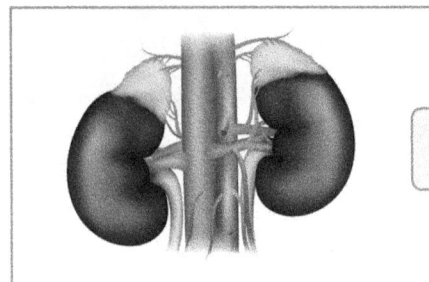

kidneys

This pair of organs filters waste from the blood and water from the body. They get rid of it as urine.

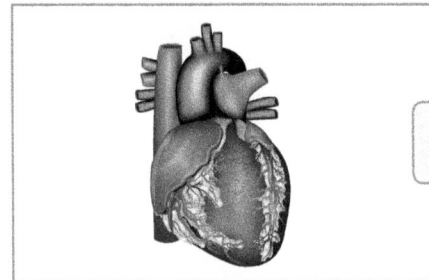

heart

This organ is like a pair of air filled bags. You need them to breathe.

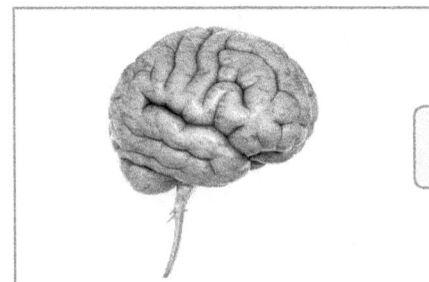

brain

This organ takes poisons out of the blood. It is the largest organ inside the human body.

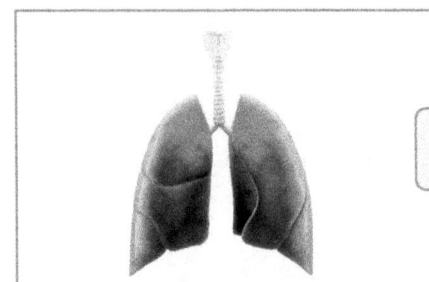

lungs

OTM2160   ISBN: 9781487710224
© On The Mark Press

Name:

Hard at Work

Can you tell where parts of your body are on or in your body? Can you tell what their job is for your body? Give it a try!

Complete the chart.

Body Part	Where is it?	What does it do?
eyes	-in my head	-helps me to see things
lungs		
teeth		
ears		
fingers		

OTM2160 ISBN: 9781487710224
© On The Mark Press

Continue to complete the chart.

Body Part	Where is it?	What does it do?
tongue		
heart		
knees		
hair		
brain		
nose		

OTM2160   ISBN: 9781487710224
© On The Mark Press